T0139080

Advancing Computational Intelligence Techniques for Security Systems Design

Security systems have become an integral part of the building and large complex setups, and intervention of the computational intelligence (CI) paradigm plays an important role in security system architecture. This book covers both theoretical contributions and practical applications in security system design by applying the Internet of Things (IoT) and CI. It further explains the application of IoT in the design of modern security systems and how IoT blended with computational intelligence can make any security system improved and realizable.

Key features:

- Focuses on the computational intelligence techniques of security system design
- Covers applications and algorithms of discussed computational intelligence techniques
- Includes convergence-based and enterprise integrated security systems with their applications
- Explains emerging laws, policies, and tools affecting the landscape of cyber security
- Discusses application of sensors toward the design of security systems

This book will be useful for graduate students and researchers in electrical, computer engineering, security system design and engineering.

Computational Intelligence Techniques

Series Editor: Vishal Jain

The objective of this series is to provide researchers with a platform to present state of the art innovations, research, and design and implement methodological and algorithmic solutions to data processing problems, designing and analyzing evolving trends in health informatics and computer-aided diagnosis. This series provides support and aid to researchers involved in designing decision support systems that will permit societal acceptance of ambient intelligence. The overall goal of this series is to present the latest snapshot of ongoing research as well as to shed further light on future directions in this space. The series presents novel technical studies as well as position and vision papers comprising hypothetical/speculative scenarios. The book series seeks to compile all aspects of computational intelligence techniques, from fundamental principles to current advanced concepts. For this series, we invite researchers, academicians and professionals to contribute, expressing their ideas and research in the application of intelligent techniques to the field of engineering in handbook, reference, or monograph volumes.

Computational Intelligence Techniques and Their Applications to Software Engineering Problems
Ankita Bansal, Abha Jain, Sarika Jain, Vishal Jain, Ankur Choudhary

Smart Computational Intelligence in Biomedical and Health Informatics
Amit Kumar Manocha, Mandeep Singh, Shruti Jain, Vishal Jain

Data Driven Decision Making Using Analytics
Parul Gandhi, Surbhi Bhatia and Kapal Dev

Smart Computing and Self-Adaptive Systems
Simar Preet Singh, Arun Solanki, Anju Sharma, Zdzislaw Polkowski and Rajesh Kumar

Advancing Computational Intelligence Techniques for Security Systems Design
Uzzal Sharma, Parmanand Astya, Anupam Baliyan, Salah-ddine Krit, Vishal Jain and Mohammad Zubair Kha

For more information about this series, please visit: https://www.routledge.com/Computational-Intelligence-Techniques/book-series/CIT

Advancing Computational Intelligence Techniques for Security Systems Design

Edited by
Uzzal Sharma, Parmanand Astya, Anupam Baliyan, Salah-ddine Krit, Vishal Jain and Mohammad Zubair Khan

CRC Press is an imprint of the
Taylor & Francis Group, an informa business

First edition published 2023
by CRC Press
6000 Broken Sound Parkway NW, Suite 300, Boca Raton, FL 33487-2742

and by CRC Press
4 Park Square, Milton Park, Abingdon, Oxon, OX14 4RN

CRC Press is an imprint of Taylor & Francis Group, LLC

ISBN: 978-1-032-13527-4 (hbk)
ISBN: 978-1-032-13528-1 (pbk)
ISBN: 978-1-003-22970-4 (ebk)

DOI: 10.1201/9781003229704

Typeset in Times
by Deanta Global Publishing Services, Chennai, India

Contents

Preface

The security system in the buildings and large complexes has become a very important issue. Even domestic households are at constant threat of theft and robbery. Because of the regularly increasing threats and the intelligence of criminals, it has gradually become very important to install such systems that can protect premises. Similarly, computational intelligence (CI) is a paradigm that is a biologically and linguistically motivated approach to establish the theory, design, application, and development of security. This book describes the application of IoT in the monitoring of entire security systems, which can provide new dimensions to systems. The application of IoT toward the design of modern security systems will be discussed with a practical approach, which will also give emphasis to research in the field of data science, which will be the main focus of the book. Overall, the book will be a handy reference for security system design. By keeping all these things in mind, proper planning and in-depth study and research are needed, which is elaborately discussed in the book. The book will give an in-depth understanding of how IoT blends with computational intelligence to improved security systems. For this book, we invited researchers, academicians, and professionals to contribute chapters, expressing their ideas and research in the application of CI and IoT in the field of security system design.

Editors

Uzzal Sharma is currently an Assistant Professor (Selection Grade) in the Department of Computer Applications, School of Technology, Assam Don Bosco University, Guwahati, India. His areas of specialization include speech signal processing, computer programming, information security and machine learning. Dr. Sharma has published more than 25 research papers in international and national journals and conferences and more than 12 book chapters in edited books. As a sole author, he has five books to his credit. He has published a Design Patent in the year 2019. He is also a member of many academic establishments and actively contributes to academics and scholarly activities. Dr. Sharma has also successfully guided a Ph.D. scholars toward their Ph.D. degree.

Parmanand Astya, Ph.D., is affiliated with Sharda University, Greater Noida, India. He has more than 27 years of experience both in industry and academia. He has published more than 150 papers in peer-reviewed international and national journals and conferences. He has also published a number of book chapters in reputed publications. He has reviewed books for publications like Tata McGraw-Hill, Galgotias Publications, etc., and papers in international journals. He had successfully completed government-funded projects and spearheaded several conferences, such as the IEEE International Conferences on Computing, Communication and Automation (ICCCA), Technovation Hackathon 2019, Technovation Hackathon 2020, International Conference on Computing, Communication, and Intelligent Systems (ICCCIS-2021), as well as conferences of IEEE student chapters. He has delivered many invited and keynote talks at international and national conferences, workshops, and seminars in India and abroad. Dr. Nand is a member of IEEE and other organizations as well as an advisory/technical program committee member of international and national conferences. He is a reviewer for a number of international journals. He has received various awards, including the best teacher award from Union Minister, a best student project guide award from Microsoft in 2015, and the best faculty award from Cognizant in 2016. Dr. Nand holds a Ph.D. in Computer Science and Engineering from IIT Roorkee, India, and M.Tech. and B.Tech. degrees in Computer Science and Engineering from IIT Delhi, India.

Anupam Baliyan is a Professor at the University Institute of Engineering and Technology, Chitkara University, Punjab. He has more than 20 Years of experience in academia. He has a MCA from Gurukul Kangari University; M.Tech. (CSE) and Ph.D. (CSE) from Banasthali University. He has published more than 30 Research papers in various international journals indexed in Scopus and ESI. He is a lifetime member of CSI and ISTE. He has been chaired many sessions at International Conferences across India. He also published edited books and chapters. He was also the assistant editor of several Scopus-indexed journals. His research areas are algorithms, machine learning, wireless networks, and AI.

Salah-ddine Krit is an Associate Professor at the Polydisciplinary Faculty of Ouarzazate, Ibn Zohr University, Agadir, Morocco, Dr. Krit is currently Director of the Engineering Science and Energies Laboratory and Chief of the Department of Mathematics, Informatics and Management. Dr. Krit received their Ph.D. degree in Software Engineering from Sidi Mohammed Ben Abdellah University, Fez, Morocco, in 2004 and 2009, respectively. From 2002 to 2008, he worked as an engineer team leader in audio and power management integrated circuits (ICs) research, design, simulation and layout of analog and digital blocks dedicated for mobile phone, and satellite communication systems using Cadence, Eldo, Orcad, VHDL-AMS technology. Dr. Krit has authored/co-authored over 130 journal articles, conference proceedings and book chapters. His research interests include wireless sensor networks, network security, smart-homes, smart-cities, IoT, business intelligent, big data, digital money, microelectronics, and renewable energies.

Vishal Jain, Ph.D., is an Associate Professor at the Department of Computer Science and Engineering, School of Engineering and Technology, Sharda University, Greater Noida, India. Before that, he worked for several years as an Associate Professor at Bharati Vidyapeeth's Institute of Computer Applications and Management (BVICAM), New Delhi. He has more than 14 years of experience in academia. He has earned several degrees: Ph.D. (CSE), M.Tech. (CSE), MBA (HR), MCA, MCP, and CCNA. He has more than 500 research citations with Google Scholar (h-index score 9 and i-10 index 9) and has authored more than 85 research papers in professional journals and conferences. He has authored and edited more than 15 books with various reputed publishers, including Springer, Apple Academic Press, CRC, Taylor and Francis Group, Scrivener, Wiley, Emerald, and IGI-Global. His research areas include information retrieval, semantic web, ontology engineering, data mining, ad hoc networks, and sensor networks. He received a young active member award for the year 2012–2013 from the Computer Society of India, and best faculty award for the year 2017 and the best researcher award for the year 2019 from BVICAM, New Delhi.

Mohammad Zubair Khan received their Master's and Ph.D. degrees in Computer Science and Information Technology from the Faculty of Engineering, M. J. P. Rohilkhand University, Bareilly, India. He was the head and Associate Professor in the Department of Computer Science and Engineering, Invertis University, Bareilly. He has more than 15 years of teaching and research experience. He is currently an Associate Professor with the Department of Computer Science, Taibah University. He has published more than 70 journal and conference papers. His current research interests include data mining, big data, parallel and distributed computing, theory of computations, and computer networks. He has been a member of the Computer Society of India since 2004.

Contributors

D. L. Kavya Reddy: Sharda University, Greater Noida

D. R. Soumya: Sharda University, Greater Noida

Gyanesh: Sharda University, Greater Noida

Subrata Sahana: Sharda University, Greater Noida

Nitin Rakesh: Sharda University, Greater Noida

Maushumi Lahon: Assam Engineering Institute

Uzzal Sharma: Assam Don Bosco University

Keshav Kaushik: University Of Petroleum And Energy Studies, Dehradun, Uttarakhand

Arti Saxena: Fet, Mriirs

Falak Bhardwaj: Fet, Mriirs

Shyam Raj S.: Fet, Mriirs

S. Kannadhasan: Cheran College Of Engineering

R. Nagarajan: Gnanamani College Of Technology

M. Shanmuganantham: Tamilnadu Government Polytechnic College

Dilip Kumar Dalei: Drdo, Bengaluru

Debasis Gountia: Cet, Bhubaneswar

Osheen Oberoi: School of Management Studies, Punjabi University, Patiala

Sahil Raj: School of Management Studies, Punjabi University, Patiala

Viput Ongsakul: National Institute of Development Administration (NIDA), Thailand

Vishal Goyal: Department of Computer Science, Punjabi University, Patiala

Shahbaz Ahmad Khanday: Department of Computer Science Engineering, Sharda University, Greater Noida, India

Hoor Fatim: Department of Computer Science Engineering, Sharda University, Greater Noida, Sharda University

Nitin Rakesh: Department of Computer Science Engineering, Sharda University, Greater Noida, Sharda University

Ragini Karwayun: Dept. of IT, Ipec, Gzb

Monika Sainger: Dept. of IT, Ipec, Gzb

Dhirendra Siddharth: Department of Computer Sciences and Engineering, Faculty of Engineering and Technology, Rama University, Uttar Pradesh, Kanpur, India

Priti Singh: Department of Computer Sciences and Engineering, Faculty of Engineering and Technology, Rama University, Uttar Pradesh, Kanpur, India

Dilip Kumar J. Saini: Department of Computer and Information Sciences, Himalayan School of Science and Technology, Swami Rama Himalayan University (SRHU), Uttarakhand, India

1 Analysis of Various Security Defense Frameworks in Different Application Areas of Cyber-Physical Systems

D. L. Kavya Reddy, D. R. Soumya, Gyanesh, Subrata Sahana, and Nitin Rakesh

CONTENTS

DOI: 10.1201/9781003229704-1

1.1 INTRODUCTION

Cyber-physical systems (CPSs) are a new class of systems designed to provide close interaction between cyber objects and physical objects. The CPS sector has been identified as a key research area, and CPSs are expected to play a key role in developing future programs [1].

While working in a CPS program center, create correspondence models that can dependably coordinate time and reaction control within the model. Cyber-physical systems screen real cycles and change how they work to assist the physical climate to work better and better [2]. A CPS contains two critical parts – actual cycles and a digital framework. For the most part, digital frameworks are an organization of little gadgets that detect calculation and communication capacities for screening actual cycles [2] (Figure 1.1).

CPSs have various applications, such as weather forecasting, aircraft control, transportation systems, civil infrastructure monitoring, deep-sea drilling, etc. CPSs have numerous highlights, like empowering singular segments to work together and delivering complex systems [3]. The actual items are outfitted with infrared sensors, scanner tags, or RFID labels, which can be filtered by smart gadgets [4]. These gadgets can be connected to the internet to send data to observe and manage the real-world situation [5].

Even though it is not currently possible to provide a conventional definition, cyber-physical systems are to a great extent alluded to as heralding a new age of systems that incorporate calculation, communication, and control to accomplish efficiency, high performance, stability, and durability as they work with physical systems [6, 7]. Cyber-physical systems are currently being broadly incorporated into different critical infrastructures due to a lack of compact mechanisms [7].

CPSs have complex security infrastructures that cause weaknesses and potential security dangers and they can be powerless against different cyberattacks while showing no indication of component failure [8].

1.2 HISTORY AND INTERPRETATION

CPSs are a new and evolving field that grants us access to new and unique connected components. In the United States, the phrase "cyber-physical system" was coined by the National Science Foundation (NSF) in 2006. As per a PCAST 2007 report, the cyber-physical system approach is now viewed as a crucial and unavoidable development toward the future of data association and data advancement (NIT). PCAST

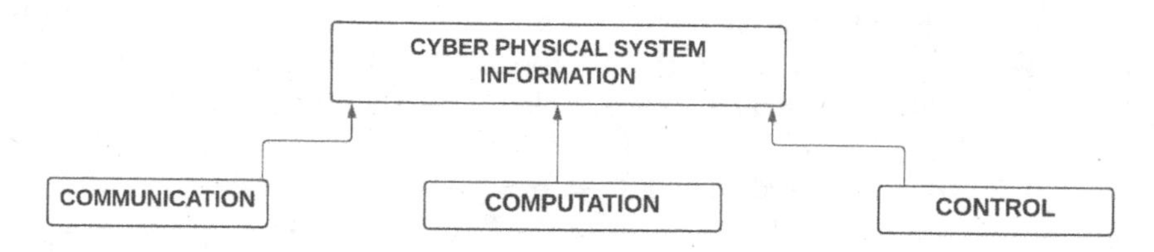

FIGURE 1.1 The 3 C concept of the cyber-physical system.

recommends revamping public services through NIT research and development (R&D) and putting CPSs at the most elevated level of the assessment plan [9, 10].

The United States set up the Digital-Physical Systems Virtual Organization (CPS-VO) to empower collaboration between CPS experts in academia, government, and business [11].

Advanced Research & Technology for Embedded Intelligence Systems (ARTEMIS), a European Union joint innovation drive, placed assets under examination and development (R&D) for state-of-the-art planning structures within public and private relationships between the European Nations and business to achieve a vision of the world in which all systems, hardware, and their components are mindful [12].

Furthermore, Horizon 2020, another investigation and research program, was launched by the European Commission (EC) around the end of 2013 to develop new procedures, ideas, and considerations for settling social difficulties. It was an ambitious and inventive program with a spending plan of EUR 80 billion. It covers CPSs through continued investigation and improvement [13].

1.3 CPS ARCHITECTURE

- The first layer is the perception layer, also referred to as the popularity layer or sensor layer (Figure 1.2). This layer has one or multiple terminal instrumentalities like actuators, sensors, international position systems (GPS), cameras, optical maser scanners, RFID tags with 2D barcode labels, and readers [12].
- The second layer is the transmission layer, conjointly referred to as the network layer or transport layer, which is additionally accountable for

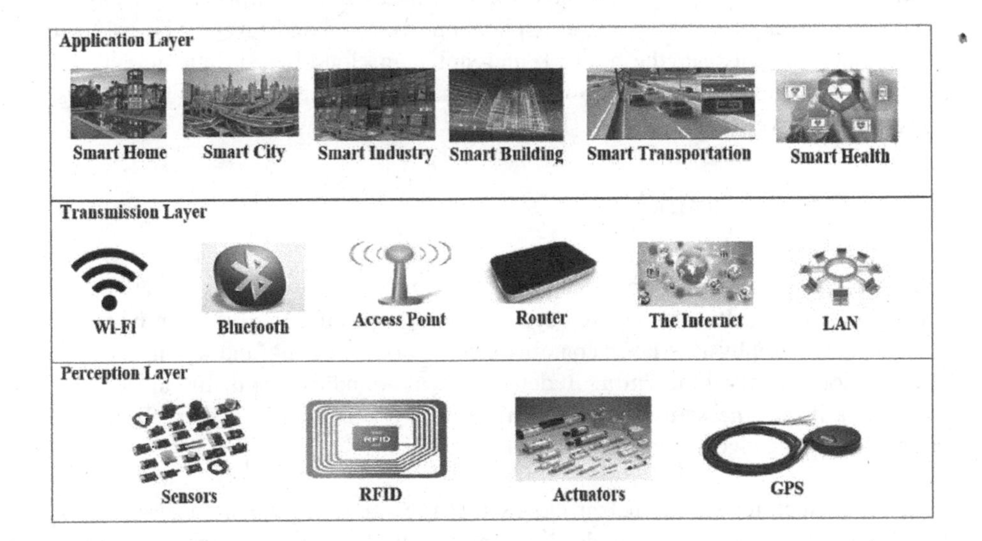

FIGURE 1.2 The layers of cyber-physical systems.

processes and interchanging knowledge between the two applications and therefore the perception layer [14] (Figure 1.2).

- The application layer is the third and most interactive (communicative) layer (Figure 1.2). It has a specific goal: it analyzes data received from the information transmission level [15].

1.3.1 Application Layer

The application layer is the last layer of the Open System Interconnection model. The application layer interface directly interacts with the applications and presents common net application services; the applied layer may connect to the presentation layer [16].

- Application layer functions include:
- Transport access and management.
- It permits the user to retrieve, access, and manage files on a remote laptop mail service.
- It provides email forwarding and storage facilities.

Other functions

- In addition to the three functions above, there are other functions: directory services, remote job entry, graphics, data communication, and so on.
- Other functions of the application layer: DNS (Domain Name System).

1.3.2 Transmission Layer

The transmission layer is responsible for transmitting data/information from the perception layer (first layer) to the application layer (second layer). The gathered information is sent over the network and stored in cloud-based data storage. IoT systems use a combination of mobile, internet satellite, communication, and wireless networks, e.g., Wi-Fi and ZigBee [17].

1.3.3 Perception Layer

The perception layer is the physical layer that senses and aggregates data regarding the encompassing. It senses physical/identified and alternative useful objects within the encompassing. Its options are used for the process and transmission of sensor knowledge. The physical layer contains sensors for detecting and assembling the information regarding the setting. It detects actual boundaries or distinguishes elective items inside the setting. The organization layer associates elective beneficial items, network gadgets, and workers [18].

GPS – Satellites carry nuclear clocks that give amazingly exact times. The time data is put in the coded broadcast of the satellite so a recipient can consistently see the time the sign was communicated.

RFID – GPS location coordinates can be sent as part of the standard beacon payload by active RFID tags with embedded GPS sensors. This tag can be read by an active RFID reader or by directly sending a beacon to a satellite.

Actuator: An actuator is a machine component that moves and controls a mechanism or system by opening a valve. It's a "mover" in basic terms. When an actuator gets a control signal, it converts the energy from the source into a mechanical motion [19].

1.4 CYBER-PHYSICAL SYSTEM CHARACTERISTICS

- Security and Privacy – CPSs are very vulnerable either due to their place in open environments or because they often communicate wirelessly, guaranteeing that such systems require close attention to safety and privacy risks, and finding techniques to shield the data is going to be necessary.
- Interoperability – Tiny parts can also be operated by completely different components on a massive scale. This allows for an essential cycle-per-second ability among heterogeneous parts and systems.
- Reliability and Dependability – Many cyber-physical systems are used for processes in our daily lives, and their utility needs high dependability and responsibility. New problems arise due to several of the CPS devices having restricted procedure power, energy, and memory.
- Power and Energy Management – The size, and therefore the autonomous operations of a number of the CPS parts, make energy management vital engineering, and therefore a design priority.
- Safety – With the proliferation of CPSs in our daily lives, it becomes extremely necessary to ensure that the actions taken on humans are safe and that risks related to these actions are identified and managed [20].

1.5 APPLICATIONS OF CPS:

- Green Buildings – By applying an integrated wireless sensing element network, the cognition manager and management systems can achieve zero net energy. Figure 1.3 shows how a wireless network operates in an existing locality.
- Smart Grid – This is an ecosystem that depends on information acquisition monitoring and the making of decisions as well as management. Figure 1.4 shows the functioning of a smart grid ecosystem.
- Medical CPS – Through the collection of diagnostic information, we can monitor patients' health and drug-related issues. Figure 1.5 shows how a CPS works in the healthcare profession [21].
- Intelligent Transportation Systems – This is a way to improve existing traffic-system control performance by constructing environments that exist in natural and manmade environments. Figure 1.6 shows various methods for controlling and organizing the transportation system. Shrewd transportation management through continuous data sharing can further develop

FIGURE 1.3 Wireless networking in our environment.

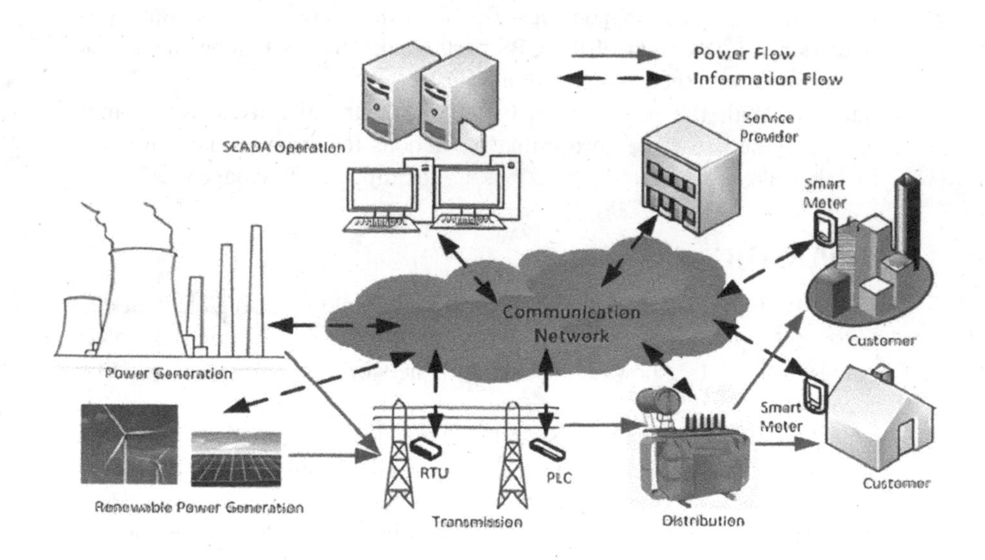

FIGURE 1.4 Smart grid ecosystem.

security, throughput, coordination, and administration in executive functions by utilizing the high-level innovations of detection, communication, calculation, and control components [22].

The achievement of zero fatalities for driverless transport, from private vehicles to public transport, centers around estimations in transportation

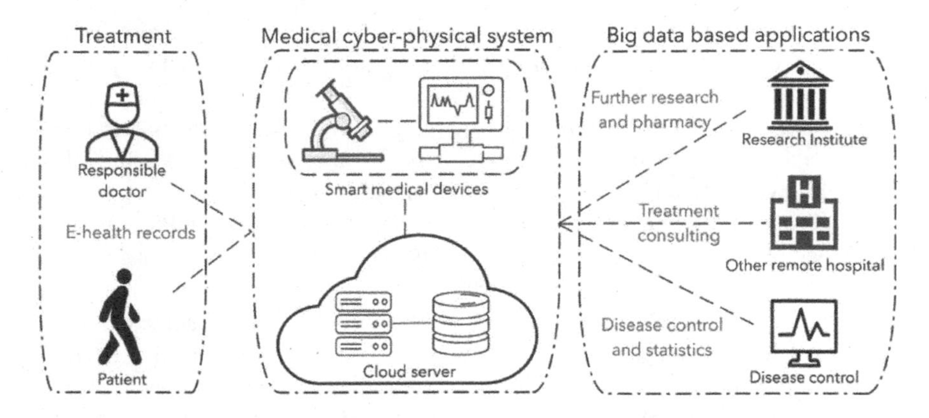

FIGURE 1.5 CPS in a medical firm.

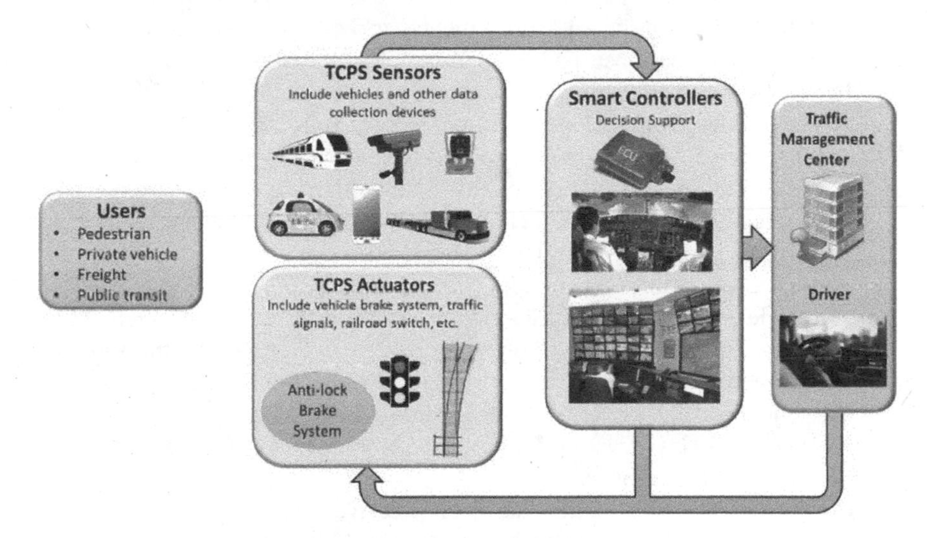

FIGURE 1.6 Types of intelligent transportation systems.

design. They count the current number of vehicles going from one geological area to another by exploiting capacities given by accurate digital maps of the actual street frameworks, which provides sequenced information for two CPS stages and frameworks, including a smart model stage for numerous automobiles with WSN routes and a digital transportation framework. They contend that CPSs advance M2M by presenting more savvy and intelligent assignments under the combination design of IoT with cyber–actual vehicle frameworks (CPVSs). They examine the time varying examination designs, sensor booking, input planning, assignment and movement arranging, as well as asset sharing and consider digital and actual assets when co-streamlining plans with portable mechanical and vehicle frameworks.

For the effective administration of a smart vehicle framework, examine the course determination of paramedics with the help of a CPS. Through the base Steiner tree approach, this assessment can ensure more and faster routes for paramedics with very low delay, minimal expense, and very high accuracy.

Vehicular digital actual frameworks (VCPS) have arisen as possible advancements for cost-effective types of assistance to end users with little delay even with a high variation of end users [23].

Propose an approach for information distribution in the VCPS climate, based on possible alliance game. Vehicles in the alliance game are accepted as the main players of the game, which get a negative (–) number of assets from the cloud-based systems. In light of the numerical establishment of the calculation for information mining, two scientists proposed a further developed methodology of a disparity measure to define the uniqueness degree between two essential tasks. This methodology can effectively tackle the issue of sensor information combinations in VCPS. Merkuryev et al. illustrate transportation, board arrangements, acknowledgment measures, utilizing a contextual investigation of Adazi, a Latin city, to further develop the accessibility level of nearby drivers to get to and from the downtown (lower altitude areas) area. In outline, shrewd transportation frameworks depend not just on cutting-edge sensors, but also on implanted PC framework innovations and remote, cell, and satellite innovations to more readily oversee complex tracking to guarantee well-being, and broaden situational mindfulness. Exploration of CPSs, such as driverless vehicles, astute convergence frameworks, and remote correspondence frameworks for vehicle-to-vehicle (V2V), and vehicle-to-foundation (V2I) communication, can assume a significant part in resolving difficulties with insightful transportation frameworks [24].

- Humanoid Robots – Robots can be used to create humanoid nurses to perform scientific investigations and perform surgeries, etc. Figure 1.7 shows the design of humanoid robots.
- Smart Manufacturing – These days, it is perceived that eligibility, seclusion, and recon durability are the primary difficulties of assembling frameworks. Shrewd assembly alludes to the utilization of implanted programming and equipment advancements to upgrade usefulness in the production of merchandise or the conveyance of administrations [14]. CPSs are a vital innovation for achieving smart manufacturing.

FIGURE 1.7 A machine to aesthetically resemble humans.

Mechanized stockroom frameworks assume a critical part in such frameworks and are presently controlled by utilizing various leveled and concentrated control structures and ordinary computerization programming procedures. Basile et al. present preliminary results for furthering eviction, particular, and disseminated control engineering for computerized stockroom frameworks; utilizing a Function Blocks and a CPS viewpoint [15] provides a basis for digital actual item administration frameworks.

CPSs and their usage (application) in mechanical cases underline that the multidisciplinary prerequisite design (RE) of the equipment, programming, and administration aspects are critical concerns for effective and dynamic (real-time) changes to CPSs in the industry [15]. CPS research in the manufacturing industry is still at an early stage, and most studies are centered around demonstration, conceptualization, and usage designs instead of on achievement [22].

Industry 4.0 alludes to the profound incorporation of cutting-edge data advancements (like CPSs) into modern situations, arrangements, and strategies. CPSs' impact on transforming the future will further develop security, efficiency, and the effectiveness of inserting interfacing framework advancements [25].

- Smart Learning Environment – To gather appropriate information about the physical environment, converting measurable data into information and knowledge will provide useful and important services for students.
- Civil Infrastructure Monitoring – For accurate and uninterrupted monitoring of infrastructures.
- Aeronautics Applications – For flight test instrumentations, pilot and crew communications, structured health monitoring. Figure 1.8 illustrates the network in an aircraft.
- Process Controls – Modern cycle control frameworks are broadly used to give self-sufficient control over creation measures through control circles. As CPSs are generally defined as combinations of calculations with actual cycles, the most immediate utilization of CPSs in Industry 4.0 situations is in improved cycle control. CPSs can give expansive powers over intricate and enormous mechanical cycles through a heterogeneous organization of the engineering of actuators, sensors, and processors, since CPSs can coordinate every one of the components to keep them synchronized. It gets conduct specifications from the underlying and data contributions of the client regarding the smart control of the actual frameworks. From the hypothesis of the Formal Concept of Analysis, Klimes examines ideas and strategies for programming framework conduct for the smart control of designed frameworks. Framework execution improvement can be accomplished by upgrading the times the regulator functions. Given the dynamic venture displaying for the huge business measure, the executives extend a current model to choose the task period in constant for CPS with the necessity for regular assignments arranged by rate calculation. With regards to CPSs, the author presented a book on control-hypothetical programming, suggesting

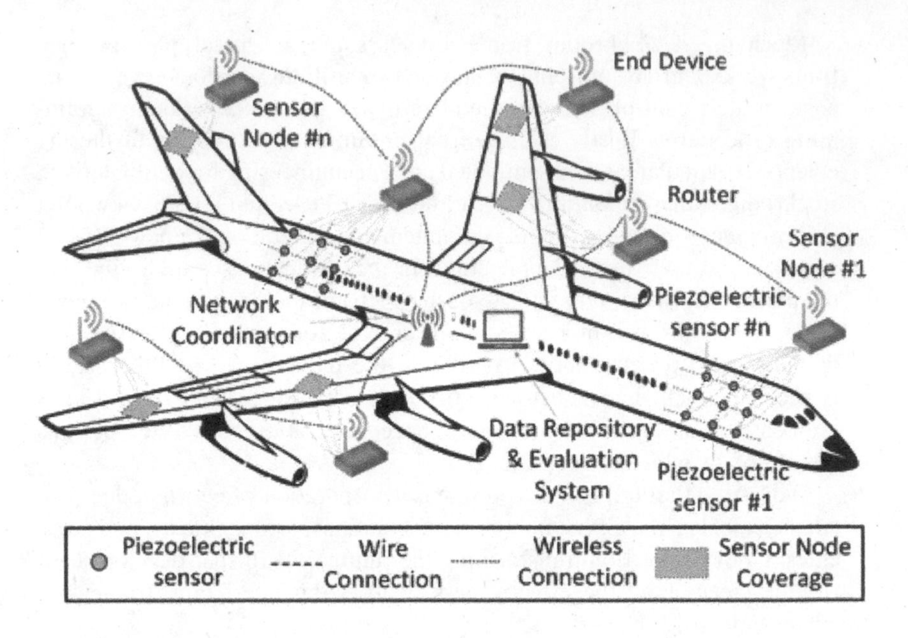

FIGURE 1.8 CPS used in aerospace.

solutions for organizing time consistency and memory use in the runtime (dynamic) checking of frameworks associated with the real world. Mr. Han proposed a framework to screen the development and direction of support during lifting on a cross-ocean [26].

1.6 SECURITY

CPSs can utilize a large amount of data from the actual climate to give something like universal, energy-efficient, and minimal expense functionality systems, and can utilize this in difficult applications, and therefore the machines (like actuators) can straightforwardly influence the real world. In this way, security is one of the current issues that needs to be examined in CPSs and thus recently this has been given considerable attention. Some scientists have provided a hypothetical structure for actual digital corporations, engaging CPS analysts to methodically plan answers to guarantee well-being, security, or manageability. They presented a system for determining the safety of a CPS, where the conduct of the CPS is constrained by a danger model that catches – in a unified manner – the digital angles (with discrete values) and therefore the actual angles (with consistent upsides) of the CPS. Chen and Chang examine the safety challenges in M2M interchanges in remote associations of CPSs and outline the restrictions and attack issues; they identified a lot of difficulties that should be attended to when creating a secure CPS. Mo and Sinopoli examined attacks on CPSs, and revealed several frameworks with safety deficiencies along with their assessment mistakes under the assault, which gives a platform for quantitative proportioning of the strength of a framework. They likewise give a close calculation for the external estimation of accessible sets. Sridhar et al. present ongoing research and efforts toward upgrading a framework's applications and provide foundation security enhancements by classifying the conditions where the actual digital controls need to assist the matrix and, therefore, the correspondence and calculations that should to be shielded from digital assault. Axelrod presents a model that assists with determining situations that cause multiple types of danger and proposes appropriate strategies to contain and limit such danger. Boyes audits the strengths of strategies against danger in intricate CPSs and examines dependability, offering a simple definition of a reliable framework. Mohan et al. present continuing engineering that upgrades the basic safety of CPSs, no matter the existence of malware [27]. Summing up the security dangers of CPSs offers recommendations for the investigation of CPSs for valuable safety endeavors, and provides security measures and proposals for a good range of threats. Gao et al. again presented the dangers to the safety of three layers: network layer (main responsible), actual layer, and application layer; they rundown the weaknesses of CPSs in terms of the executives and strategy, stage, and organization [13]. They analyzed the data of disturbance assaults, like Stuxnet, and officially clarified that confidence in the network protection of a CPS plays into the accomplishment of the assaults. They characterize how the attacks cover-up and utilize the system administrator's faith to remain undetected (uncaught) and presume that confidence in the CPS is significant enough to achieve the assault [13]. Peisert et al. examine the importance of difficulties with designed-in security consistent with different viewpoints including a master from the scholarly community, a CPS supplier, and an end resource proprietor. Proper assessment of safety has consistently been one of the biggest difficulties within the field of PC securities. The reconciliation of calculation and communication advancements with actual segments has presented an assortment of security changes, compromising actual digital segments. Orojloo and Azgomi propose another methodology for displaying and quantifying assessments of the safety of CPSs. The offered technique provides models of the various classes of hostile attacks against CPSs. So far, there are many bugs in the framework models and the security structures of CPSs. Sun et al. provide an in-depth and a brief survey of the present writing on certainty in CPSs for future remote age frameworks, distinguishing the features of CPSs and proposing a model design with certainty; they also give further attention to the security of those frameworks . Wang and Yu break down the security model within the theoretical digital actual gaseous petrol pipeline framework hooked into Petri-net and lay a foundation for investigating confidentiality and data security in CPSs. Wan as well as Alagar overview the status of CPS security, distinguishing the problems encompassing secure control and exploring the degree to which setting data could be utilized to further develop the security and growth of CPS. Vegh and Miclea present an amazing solution for guaranteeing information privacy and security enhancements by consolidating two conventional techniques used for digital safety – the cryptography technique and steganography technique [19].

They describe the four potential results of a digital assault on a CPS machine and suggest trust prerequisites that a gadget should fulfill to ensure correct conduct function throughout the gadget's entire life. The "Trustworthy Autonomic Interface

Guardian Architecture" (TAIGA), which screens communication between an implanted regulator and the actual cycle, provides a particular cycle as the last line of security against digital attacks [18]. The authors proposed an ingenious method for creating online protection plans for actual frameworks. They contend that on both the attacker and thus the safeguard side to achieve their most prominent pay off. The perfect regulator can continue with the straight framework during a steady way when the digital state vector meets a specifically wanted basis [20]. The author reviews late examination of guarding in cloud-based (environment) CPSs and investigate the safety issues of present-day design gadgets and smart versatility administrations [21]. Present the "cyber-physical attacks description language" (CP-ADL), which creates a foundation for the organized portrayal of assaults on CPSs. CP-ADL broadens the scientific classification, depicting connections between connotation and unmistakable perspectives, notwithstanding the overwhelming connections that present assaults on the CPS.

CPSs are mind-boggling and diverse frameworks with many weaknesses. Bou-Harb examines various approaches for the security of CPSs from a control-theoretical and network safety point of view and highlights the danger identifiers in several CPS conditions. Fernandez examined how to develop misuse designs to illustrate the dangers to CPSs and how to count and bind them together to build security.

The author [26] present an investigation of verifiable security episodes to dissect the elements of a selected class of CPSs, modern-control-frameworks (ICS), and mention how key qualities of upcoming savvy CPSs in arenas such as modern settings can present further difficulties concerning handling dormant plan. Sanjab and Saad studied a general model for a CPS, which included the diffusion of an attack on the digital layer of a particular framework. This overall methodology was used to grasp network security by taking a wide region of insurance with power market suggestions. Vegh and Miclea depict a method that utilizes a progressive cryptosystem to make sure communication inside a CPS is consolidated with complex events to streamline the safety design. Wang examines the security conditions of strategic distances and their emphasis on organization in CPSs and gives another conventional strategy to research the data security properties. Wang examines the delicacy of an electrical digital actual framework (ECPS) under different cyberattacks with forswearing-of-administration (DoS) assaults, replay assaults, and bogus information infusion assaults [28].

The cyberattacks cause problems on CPS and require signing – significant social and monetary misfortunes. Making a CPS secure is always difficult thanks to the assortment of assault bottom layers from the digital and actual segments, and this frequently restricts calculation and communication between assets [12]. Propose a cross-layer plan structure for an asset compelled CPS. The structure joins control-hypothetical strategies at the sensible layer and network protection methods at the inserted stage layer, and addresses security alongside other plan measurements, for instance, control execution under asset and permanent requirements [12, 16]. Investigate the digital–physical testbed created inside the EU Project FACIES to interrupt the way screen frameworks are normally utilized in Mechanical Control Systems, which could be inclined to fail when faced with digital assaults. They then diagram how the existing presence of a digital Intrusion Detection System (IDS) works on the activity and therefore quality of the insurance blueprint [16, 28]. Direct a systematic planning study (SMS) to distinguish 48 essential Model-Based-Security-Engineering for CPS (MBSE 4PS). Their outcomes not only show that the message is the simplest in MBSE 4PS, but additionally calls attention to a couple of open issues that might merit more examination, for instance, the absence of designing security solutions for CPSs, restricted device support, barely any modern contextual analyses, and therefore the challenge of crossover DSLs in designing secure CPSs.

CPS security must vary from online protection. CPS security has a higher priority than network safety, and online protection is often considered an important part of CPS security due to the digital segments of CPS, which demands all of the cyber-physical system network protection conventions, notwithstanding the parallel security conventions presented by the particular parts and their corporations [29].

1.7 CHALLENGES AND ISSUES FACED BY CPS

Technical and non-technical problems in environmental systems are referred to as CPS challenges. Sensing, networking, power management, cloud, complexity, privacy, reliability, data management, and security are all difficulties that CPSs encounter. We observed that protection and security are significant difficulties or issues for CPSs. Another critical function of CPSs is to oversee huge amounts of information because various gadgets are associated with one another through the web, and a lot of information is created by each associated gadget. For collecting, recognizing, analyzing, and making sense of the enormous amount of data created, as well as safeguarding it from cyberattacks, a solid solution is necessary. This should be taken into consideration while looking at security issues.

The security is associated with data and control security. Data security is also known as information security, and focuses on data exchange and protection in the network, using a cryptographic mechanism to protect against cyberattacks. Control security focuses on preventing cyberattacks by safeguarding the components of control frameworks. The following security issues are discussed, along with strategies to defend against them [25].

- Security for devices – Smartphones, sensors, and other devices are among the equipment available or used in CPSs. Our purpose is to prevent unauthorized access to these devices as well as device disablement. System services, hardware resources, and data, both in transit and storage, may all be protected using computer security approaches. New and current mobile banking applications for making online payments are available on smartphone devices. Hackers have targeted these devices and taken over access to a commit fraud. For this, a robust security strategy should be used to safeguard different mobile devices.
- Secure data transmissions – In a secure communication network, detecting fakes and malicious activities, as well as preventing unauthorized access, is crucial. Data should be protected from intrusions, eavesdropping, malicious

attacks, DDoS, and unauthorized modification between the sender and the receiver devices.

- Secure data storage – We know that the data is stored on a system or server. In CPSs, data storage security is a major concern. To access system data, you will need a username and a password. Cryptographic methods can encrypt data in large-scale storage devices [30].

1.7.1 Implementation of the Task in CPS

Workflow of CPSs is broadly classified into four steps:

- Tracking – A fundamental feature of CPSs is to monitor physical processes and the environment. It's also used to provide input on any previous CPS activities and ensure proper operations in the future. The physical process aims to achieve the CPS's original physical goal.
- Networking – This step deals with data aggregation and diffusion. In CPSs, there can be more than one sensor. These sensors can deliver information progressively, and numerous sensors can produce a lot of information that should be collected or diffused before being handled by analyzers. Different programs must deal with several networking communications at the same time.
- Computing – This step involves reasoning and evaluating the data gathered during monitoring to decide if the physical process meets predetermined requirements. If the criteria are not being followed, corrective measures are recommended to ensure that the objectives are satisfied.
- Actuation – The behavior decided on during the computing process is carried out in this stage. Actuation may be used to perform various tasks, such as correcting the CPS's cyber behavior or altering physical operation. For example, in a medical CPS, the operation may be the distribution of medicine [17].

1.8 RISK ASSESSMENT

With the expanded utilization of CPSs in numerous sensitive areas (e.g., healthcare and smart housing), security has become a critical issue, and there is a requirement for a suitable danger appraisal approach.

Since CPS security is distinct from standard IT systems, the security features are likewise distinct; for instance, they include unsecured connectivity and standardized protocols and technologies. The point of CPS security assessment is to have an calculated type of risk that can be utilized in future framework assurance. The vast majority of studies that focus on venture frameworks are not straightforwardly identified with CPS [15]. Since CPS security is different than for a conventional IT system, the security features are likewise contrasting, for instance, they include normalized conventions and the advancements and unreliable interconnections [15].

1.8.1 Asset Identification

A resource, which alludes to resource esteem that should be secured [17], can be solid (for example, clinical gadgets and so on) or intangible (e.g., data about an organization). Indeed, most resources are immaterial in this manner, and resources can have an immediate benefit to numerous day-by-day exchanges and administrations.

Resource quantification can be estimated by immediate and indirect financial misfortune and the outcomes of harm. The value appraisal measure includes the ID of the protection layers, basic resources, and the fundamental elements of the framework, just as when evaluating the resource value rating. CPS resources can likewise be partitioned into three parts: actual resources, cyber resources, and intuitive systems. The fundamental contrast between CPS resources and traditional IT resources is that the inward communication of CPSs are difficult to understand, immaterial and contain interconnections with different frameworks.

1.8.2 Threat Identification

This development is likewise used to help recognize hazards in high need of concern in the field of CPSs, which is anything but a simple task. Past information can be utilized for measuring the recurrence of danger while examining records and logs in Intrusion Detection Systems (IDS) and can be used to predict the recurrence of danger to logs and many other techniques. Mitchell and Chen introduced an extensive study that characterized a new CPS IDS procedure, presented research directions, and additionally summed up the most examined CPS IDS strategy in the field [12].

1.8.3 Vulnerability Identification

Vulnerability can be characterized as any current shortcomings or imperfections that could be misused for spying or criminal purposes by a antagonist to hurt the worth of a resource. It can likewise be characterized as the condition or climate that can be abused by an enemy to harm, ruin, attack, or destroy systems. A weakness evaluation is the examination of a framework and its functioning, distinguishing shortcomings or blemishes or breaks, and deciding best remedial action to lessen or remedy any weaknesses.

CPS weaknesses are regularly separated into three parts: organization, stage, and board. Organization weakness incorporates arrangement, equipment, and observation weaknesses. Stage weakness includes setup, equipment, and programming weaknesses as well as imperfections in security measures. The board weakness is often identified with the absence of a safety strategy. Weakness quantification can be determined through alternate systems like the past master assessment technique, comparing the authentic records of enterprises [17]. Eliminating or forestalling all danger is a difficult task, if not near impossible. In a like manner, most minimal expense strategies can be typically taken to diminish the danger to an adequate level.

1.9 CYBER-PHYSICAL SYSTEM SECURITY

The different capacities of CPSs assist with crucial mechanical processes. Any security relaxation will lead to very serious consequences. Cyber-physical systems can screen embedded physical processes, so compromised information will result in the loss of confidentiality and potential harm to sensitive information. Cyber-physical systems are utilized in power management, so any malignant action will result in attacks. It has the potential to have an impact on the environment. Unauthorized movement will bring damage to the interaction [10].

1.9.1 CPS Security Requirements

- Privacy – In CPSs, a large number of different information measures are continually occurring, and this is something that the vast majority of users don't know about [6, 7]. As a result, an individual has the advantage of accessing their data and understanding what kind of data is being gathered about them by data gatherers and to whom these data are being supplied or offered. Furthermore, this requires thwarting the unlawful/unapproved access to the customer's data and disclosure of their data [8].
- Dependability – The Intelligent Physical World makes certain that flexible CPS performance can deliver high reliability and guarantee quality-of-service through its adaptation to non-critical failure systems in an opportune way. Steadfastness incorporates two different characteristics, well-being and dependability. Security is regularly characterized by the association's objectives [29].
- Coordination and Interaction – These are fundamental to keeping CPS security operational. The communication and coordination of cyber and physical framework components are a crucial aspects of cyber-physical systems [9].

1.9.2 Cyber-Physical Systems Security Challenges

The selection of safety measures has numerous advantages with regard to securing CPS segments, layers, and areas. Regardless, despite these amazing advantages, CPSs are influenced by the usage of security endeavors, which can be described as follows:

- Lesser Performance – Well-being endeavors can for the most part impact the function of a CPS, without a good idea of a reasonable security–execution risk. This can impact conventional assignments and requires more human mediations to direct administrations and areas.
- Operational Security Delays – Upon the arrangement of any security management, there is a preparation stage that goes before the fully operational security mode, and during which security is briefly inadequate or rudimentary and, in this manner, inclined to assaults.

- Expensive – Higher/advanced security levels are related to higher/advanced computational expenses, which are not restricted to the underlying capital spending stage, but incorporate the preparation, update, and operational stages.
- Compatibility Issues – Many CPS frameworks are incompatible with the safety systems in place and vice versa. This might be due to the software in use, such as the firmware or the operating system.
- High Power Consumption (HPC) – This is indeed a significant issue, especially both for asset-driven and battery-restricted CPS end devices. A more powerful utilization implies a more limited life expectancy and a greater expense to maintain operation.

1.10 CYBER-PHYSICAL SYSTEMS SECURITY SOLUTIONS

Maintaining a safe CPS environment is certainly not a simple task because of the consistent increment of difficulties, incorporation issues, and restrictions of the current arrangements, including the absence of safety, protection, and accuracy. Table 1.1 displays a classified table of cyber-physical breaches and security measures performed in various components. In any instance, numerous approaches, including cryptographic and non-cryptographic plans, can be used to accomplish this.

TABLE 1.1
CPS Security Goals

Security Goals	Physical Breach	Physical Measures	Cyber Breach	Cyber Measures
Privacy	Social Engineering Phishing Blackmailing	Employee training Raising Awareness, Legal Measures	Malware- Virus Trojan	Anti-Malware
Confidentiality	Stealing CPS documents	Document Classification	Eavesdropping Wiretapping	Lightweight Cryptography Algorithms
Integrity	Social Engineering Breach of Agreements	Legally/Signed Agreements	Malicious/False Data Data Manipulation	Anti-Malware
Availability	Physical Device Damage/ Destruction	Tamper Resistant Devices	Denial of Service Signal Jamming	Firewalls Backup Devices
Authentication	Abuse of privilege	Employee Screening Accountability	Weak password Recovery Password Cracking	Strong Password Mechanism

1.10.1 Solutions Based on Cryptography

Cryptographic measures are fundamentally utilized to protect the channel of communication from dynamic or inactive attack, along with any unapproved access and obstruction, especially in SCADA systems. Truth be told, customary cryptography approaches are dependent on using codes and hash cannot be easily applied to CPSs, including IoCPT, because of force and size limitations. Thus, the fundamental center should be restricted to information security alone; rather, it must only maintain and guarantee the effectiveness of the general framework measure.

1.10.1.1 Security Goals of Cyber-Physical Systems

- Confidentiality means an asset that only allows authorized groups to access sensitive data created within the framework. Data storage is a significant issue that should be tended to in many CPS frameworks. For instance, in a crisis, the tactile board organization that discerns the security of the data sent may reduce the risk of a crisis in the executive framework. The classification of information communicated by the sensor hubs might be undermined and may cause the progression of information over the organization of the affected sensors, sensitive data being scaled down, or fake nodes performing on the network. In addition, false/harmful data can be added to the network in addition to those fake nodes. Therefore, the privacy of the data distribution needs to be kept at a certain level.
- Integrity means the assets of a defense system or the information contained therein that prevents unauthorized fraud or alteration to maintain the correct information. The high-fidelity system should provide comprehensive accreditation and consistent testing procedures. High integrity is one of the critical highlights of cyber-physical systems.
- Availability means that for any framework to fill its need, the service should be accessible when it is required. It implies that the digital frameworks used to store and handle the data, the actual controls used to perform actual cycles, and the correspondence channels used to get to it should be working effectively. The high accessibility of CPSs should consistently offer support by forestalling calculations, controls, and communications because of equipment failures, framework overhauls, or forswearing-of-administration assaults [2].
- Authenticity is that, with registering and communication, it is important to guarantee that the information, exchanges, and interchanges are real. Likewise, it is significant for safety to confirm that the two players included are whom they say they are [5]. In CPSs, authenticity includes verification of all the connected processes like actuation, sensing, and communication.

1.10.2 Solutions Based on Non-Cryptography

Numerous non-cryptographic solutions have likewise been introduced to relieve and remove any conceivable digital attack or malicious event. This was done by executing honeypots, firewalls, and Intrusion Detection Systems.

- Firewalls – Because of the invention of IDS and AI technologies, firewalls have rarely been used in the CPS sector. As a result, only a few firewall-based solutions have been proposed, which use paired firewalls between the corporate and industrial zones to improve server cybersecurity. 23They choose paired firewalls because of the strict security and clear administrative separation [23].

REFERENCES

1. L. Monostori et al., "Cyber-physical systems in manufacturing," *Cirp. Ann.*, vol. 65, no. 2, pp. 621–641, 2016.
2. E. K. Wang, Y. Ye, X. Xu, S.-M. Yiu, L. C. K. Hui, and K.-P. Chow, "Security issues and challenges for cyber physical system," in *2010 IEEE/ACM international conference on green computing and communications \& international conference on cyber, physical and social computing*, 2010, pp. 733–738.
3. L. Vegh, and L. Miclea, "Enhancing security in cyber-physical systems through cryptographic and steganographic techniques," in *2014 IEEE international conference on automation, quality and testing, robotics*, 2014, pp. 1–6.
4. R. Khan, S. U. Khan, R. Zaheer, and S. Khan, "Future internet: the internet of things architecture, possible applications and key challenges," in *2012 10th international conference on frontiers of information technology*, 2012, pp. 257–260.
5. B. Zhang, X.-X. Ma, and Z.-G. Qin, "Security architecture on the trusting internet of things," *J. Electron. Sci. Technol.*, vol. 9, no. 4, pp. 364–367, 2011.
6. S. Belguith, N. Kaaniche, M. Mohamed, and G. Russello, "C-ABSC: cooperative attribute based signcryption scheme for internet of things applications," in *2018 IEEE international conference on services computing (SCC)*, 2018, pp. 245–248.
7. A. O. Moyegun, *Information security and innovation; guide to secure technology innovation initiatives*, 2016.
8. N. Kaaniche, and M. Laurent, "Data security and privacy preservation in cloud storage environments based on cryptographic mechanisms," *Comput. Commun.*, vol. 111, pp. 120–141, 2017.
9. G. Loukas, D. Gan, and T. Vuong, "A review of cyber threats and defence approaches in emergency management," *Futur. Internet*, vol. 5, no. 2, pp. 205–236, 2013.
10. S. Belguith, N. Kaaniche, and G. Russello, "PU-ABE: lightweight attribute-based encryption supporting access policy update for cloud assisted IoT," in *2018 IEEE 11th international conference on cloud computing (CLOUD)*, 2018, pp. 924–927.
11. Q. Shafi, "Cyber physical systems security: a brief survey," in *2012 12th international conference on computational science and its applications*, 2012, pp. 146–150.
12. B. Zheng, P. Deng, R. Anguluri, Q. Zhu, and F. Pasqualetti, "Cross-layer codesign for secure cyber-physical systems," *IEEE Trans. Comput. Des. Integr. Circuits Syst.*, vol. 35, no. 5, pp. 699–711, 2016.
13. G. Howser, and B. McMillin, "A modal model of Stuxnet attacks on cyber-physical systems: a matter of trust," in *2014 Eighth international conference on software security and reliability (SERE)*, 2014, pp. 225–234.
14. V. Gunes, S. Peter, T. Givargis, and F. Vahid, "A survey on concepts, applications, and challenges in cyber-physical systems," *KSII Trans. Internet Inf. Syst.*, vol. 8, no. 12, pp. 4242–4268, 2014.
15. S. Wiesner, E. Marilungo, and K.-D. Thoben, "Cyber-physical product-service systems–challenges for requirements engineering," *Int. J. Autom. Technol.*, vol. 11, no. 1, pp. 17–28, 2017.

16. G. Bernieri, E. E. Miciolino, F. Pascucci, and R. Setola, "Monitoring system reaction in cyber-physical testbed under cyber-attacks," *Comput. \& Electr. Eng.*, vol. 59, pp. 86–98, 2017.
17. A. M. Gamundani, "An impact review on internet of things attacks," in *2015 international conference on emerging trends in networks and computer communications (ETNCC)*, 2015, pp. 114–118.
18. H. Niu, C. Bhowmick, and S. Jagannathan, "Attack detection and estimation for cyber-physical systems by using learning methodology," in *Artificial neural networks for engineering applications*, Elsevier, 2019, pp. 107–126.
19. K. G. Lyn, L. W. Lerner, C. J. McCarty, and C. D. Patterson, "The trustworthy autonomic interface guardian architecture for cyber-physical systems," in *2015 IEEE international conference on computer and information technology; ubiquitous computing and communications; dependable, autonomic and secure computing; pervasive intelligence and computing*, 2015, pp. 1803–1810.
20. J. Puttonen, S. O. Afolaranmi, L. G. Moctezuma, A. Lobov, and J. L. M. Lastra, "Security in cloud-based cyber-physical systems," in *2015 10th international conference on P2P, parallel, grid, cloud and internet computing (3PGCIC)*, 2015, pp. 671–676.
21. M. Yampolskiy, P. Horváth, X. D. Koutsoukos, Y. Xue, and J. Sztipanovits, "A language for describing attacks on cyber-physical systems," *Int. J. Crit. Infrastruct. Prot.*, vol. 8, pp. 40–52, 2015.
22. S. Choi, G. Kang, C. Jun, J. Y. Lee, and S. Han, "Cyber-physical systems: a case study of development for manufacturing industry," *Int. J. Comput. Appl. Technol.*, vol. 55, no. 4, pp. 289–297, 2017.
23. N. Jiang, H. Lin, Z. Yin, and C. Xi, "Research of paired industrial firewalls in defense-in-depth architecture of integrated manufacturing or production system," in *2017 IEEE international conference on information and automation (ICIA)*, 2017, pp. 523–526.
24. J. Zalewski, S. Drager, W. McKeever, and A. J. Kornecki, "Threat modeling for security assessment in cyberphysical systems," in *Proceedings of the Eighth annual cyber security and information intelligence research workshop*, 2013, pp. 1–4.
25. Y. Ashibani, and Q. H. Mahmoud, "Cyber physical systems security: analysis, challenges and solutions," *Comput. & Secur.*, vol. 68, pp. 81–97, 2017.
26. S. Frey, A. Rashid, P. Anthonysamy, M. Pinto-Albuquerque, and S. A. Naqvi, "The good, the bad and the ugly: a study of security decisions in a cyber-physical systems game," *IEEE Trans. Softw. Eng.*, vol. 45, no. 5, pp. 521–536, 2017.
27. W. Guo, and W. Zhang, "A survey on intelligent routing protocols in wireless sensor networks," *J. Netw. Comput. Appl.*, vol. 38, no. 1, pp. 185–201, 2014, https://doi.org/10.1016/j.jnca.2013.04.001.
28. P. H. Nguyen, S. Ali, and T. Yue, "Model-based security engineering for cyber-physical systems: a systematic mapping study," *Inf. Softw. Technol.*, vol. 83, pp. 116–135, 2017.
29. S. Deshmukh, B. Natarajan, and A. Pahwa, "State estimation in spatially distributed cyber-physical systems: bounds on critical measurement drop rates," in *2013 IEEE international conference on distributed computing in sensor systems*, 2013, pp. 157–164.
30. R. Mahmoud, T. Yousuf, F. Aloul, and I. Zualkernan, "Internet of things (IoT) security: current status, challenges and prospective measures," in *2015 10th international conference for internet technology and secured transactions (ICITST)*, 2015, pp. 336–341.

2 Role and Development of Security Architecture and Models in Software Systems

Maushumi Lahon and Uzzal Sharma

CONTENTS

DOI: 10.1201/9781003229704-2

2.1 INTRODUCTION

The area of user requirements is so dynamic that security measures need to adapt to the changing needs of the software systems, which places great demand on the development of security architectures and models. The different areas and security levels of hardware, protocols, network devices, operating systems, and applications need to be understood to minimize the security vulnerabilities that can affect the environment. Security models are designed to implement certain security policies which are designed to fulfill certain security requirements. Systems have many aspects that need to be secured, and security can be used at various levels. Incorporating security during the design and architectural phase can reduce the vulnerability of the system rather than incorporating security measures into the system later. The objective of security architecture is to determine the hardware, operating system, and software to design a secure computer system.

Before going into the concepts of security, it is important to understand the architecture of a computer. This will provide an understanding of the relation and working of the different parts of a computer.

- Central Processing Unit – This unit consists of the arithmetic and logic unit and the control unit, and is responsible for the processing of data. It consists of registers for holding data and can be considered like the brain of the computer.
- Memory – Holds the data, applications, operating system and firmware. Different types of memory include primary memory, secondary memory and virtual memory.
- Operating system (OS) – The OS acts as a resource manager and is responsible for ensuring that none of the user applications violates the security policy.
- Processes – Programs that execute in their own address space and communicate with other processes for completion. A process can be in any one of the following states: ready, running, waiting, and stopped.
- Protection rings – This is a security mechanism that the operating system uses to protect the critical components of the system.

All the above concepts and many more are related to security and form an integral part of security models [1].

2.2 CONCEPTS IN SECURED SYSTEM ARCHITECTURE

Security is the most vital issue for security professionals. Every organization requires protection for their data to prevent unauthorized access. The function of the security needs to be decided, whether protection is to be provided to secure the files at the file system level or to restrict the operations and activities of the users. The next point of decision making relates to the placement of the security mechanism. The issue here is to identify the layer where the security mechanism is to be implemented.

Layers separate the functionality of the hardware and software into separate modules. Though the security architecture layer names are not standardized across all architectures, the following layers form a general list [1]:

1: Hardware
2: Kernel
3: Operating System
4: Device drivers/services
5: Applications/programs

Implementation of security at different layers varies in functionality and complexity. The more the layer is closer to the user, the more complexity there is in the implementation, and the assurance is less. The reduction in assurance is due to the complexity of the implementation tools.

The next step is to build and integrate the decided upon security mechanism with other parts of the system. In this process, identification of trusted subjects and objects is required. Not all components fall within the trusted component base. The trusted component base includes hardware, software, and firmware, which do not violate the security policy. As all components need not fall under the trusted component base, a security perimeter separating the trusted and the untrusted components is to be established. The difference between the trusted and untrusted components lies in the protection mechanism incorporated in the components. Proper interfaces are required in case a trusted component has to communicate with an untrusted component so that communication between them does not bring in security compromises that are not intended.

The concepts of reference monitor and security kernel come next to ensure the subject-object access permission process. To accomplish a higher level of trust, the programs, users, or processes should be fully authorized before accessing files or other resources. The reference monitor concept serves this purpose. It mediates so that the subjects may have requisite rights to protect the objects. The reference monitor is a concept that is enforced by the security kernel. The security kernel, which comprises hardware, software, and firmware, is responsible for the interactions between the subject and object. The four main requirements of the security kernel are [1]:

- Isolate the processes that carry out the reference monitoring, so they cannot be tampered with.
- Invoking the reference monitor for every access to implement the concept in a complete manner.
- Writing every decision into an audit log and verifying for correctness.
- It must be small enough to be verifiable and testable in a complete and comprehensive manner.

Other concepts related to security include layering, data hiding and abstraction. These concepts are used to protect the subjects and objects and are the foundation of a security model. Layering separates processes and resources and provides modularity to the system. The layers communicate through interfaces. Sometimes it is not required for a process to communicate with other processes in other layers. In that case no interfaces are provided for interaction. This is termed "data hiding." To make the management of objects easier, they are classified, and specific permissions and acceptable activities for the class are defined. This is called "abstraction."

2.3 SECURITY MODELS

Security models form an important concept when designing secure systems. Security policies are implemented using security models. Security models are represented mathematically or analytically and then incorporated into system specifications. Finally programmers develop secure systems based on the specifications developed after mapping the security model. Various security models are developed based on different security policies. A brief outline of the different categories of security models and overviews of some security models are given in this section [2].

2.3.1 Categories of Security Models

2.3.1.1 State Machine Models

A state represents an instance or snapshot of a point in time. State machine models are based on the particular state of the system when there is an interaction with an object. The basic aim is to take the system from one secure state to another, thus keeping the system constantly in a secure state. The activities or events that change the state of a system should not put the system into an unsecure state. Thus this means that the system is always in a secure state, which also means that when the system fails it fails in a secure state.

2.3.1.2 Information Flow Models

In this category of models, unsecure information flow is not allowed. It not only deals with the type of information but also with the direction of the flow. The concern of the models in this category is with unsecure information flows on the same level, as well as between different levels. So the models in this category allow a flow of information between different security levels or from one object to another at the same level until any restricted operation is attempted.

2.3.1.3 Non-Interference Models

This category of models emphasizes non-interference in multilevel security. The flow of information is not a concern here, but it is concerned with what a subject knows regarding the state of the system. The basic objective of this model category is to maintain the changes made at a particular level without affecting the entities at other levels. Thus entities at lower levels should not know about the activities and commands executed at a higher level, as this could be considered as information leakage.

2.3.2 Overview of Some Security Models

2.3.2.1 Bell-LaPadula Model

In 1976 a paper published by David E. Bell and Leonard J. La Padula proposed this model. It is based on confidentiality and focuses on mandatory security mechanisms. It is a mathematical model based on a multilevel security policy. The information is classified into different levels, with top secret at one end and unclassified at the other end. This model uses subjects and objects that are at different security levels. The model proposes rules that govern the acceptable activities between them. According to the concepts in this model the flow of information is always secure. The three rules governing the model are [2]:

Simple security rule – It states that a subject cannot read data that is at a higher security level.
Property rule – It states that a subject cannot write information to a security level that is lower than the existing level of the subject.
Strong star property rule – It states that a subject with the capacity to read and write can perform the functions only at the same security level.

Figure 2.1 represents the lattice of the rights of a subject in the context of security in the LaPadula model. The LaPadula model addresses only confidentiality aspects of information and not data integrity.

2.3.2.2 Biba Model

Kenneth J. Biba developed this model in 1977. The model is the first model designed to ensure data integrity. Data integrity preservation may be considered as prohibiting data from being modified by unauthorized users and also not allowing unauthorized modification by authorized users. Data integrity is also concerned with maintaining

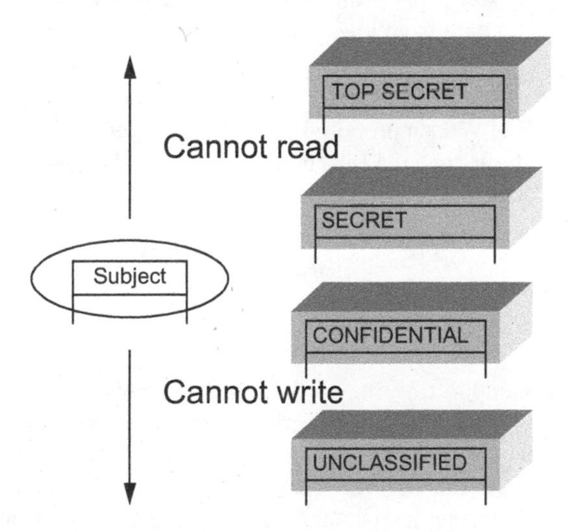

FIGURE 2.1 Rights of a subject in the Bell-LaPadula model.

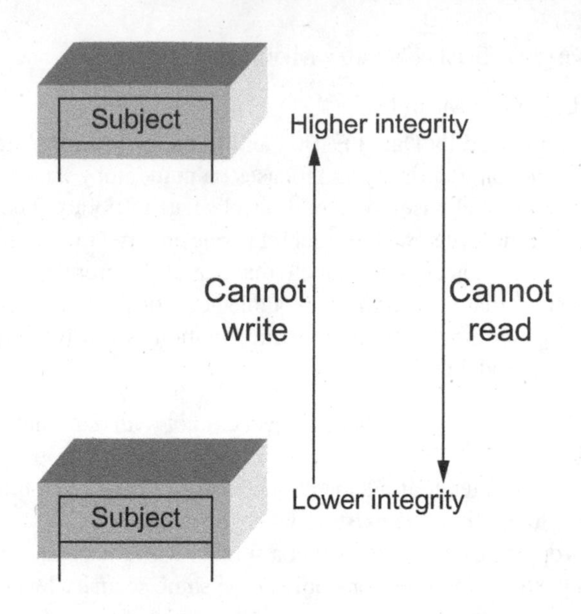

FIGURE 2.2 Biba model rules representation.

data from the real world. The model is similar to the LaPadula model but addresses data integrity not confidentiality. The model is based on two rules:

A subject cannot write data to an object which is at an upper integrity level.
A subject cannot read data from a lower integrity level.

Figure 2.2 represents the integrity rules of the Biba model. The purpose of this model is to provide integrity, i.e., it is meant to ensure that assets with high integrity levels are not corrupted with data from lower integrity levels.

2.3.2.3 Clark-Wilson Model

David Clark and David Wilson proposed this model in 1987. As the LaPadula model is based on confidentiality and the Biba model is based on integrity, the Clark-Wilson model is based on the following concerns:

All users must be properly authenticated
Modifications must be recorded
Prevent unauthorized modifications

The security policy supporting the model can be expressed using objects whose integrity is protected, objects whose integrity is not protected, and integrity verification procedures. The rule can be placed as follows:

A user if authorized can perform operations on integrity-protected data items. Operations mean modifying the data item or accepting user inputs and creating new integrity-protected data items. This ensures that the system will be in a secured state.

In other words, the Clark-Wilson model is designed to prevent authorized users from making unauthorized modifications. This is enforced by breaking up an operation into different parts. Different users then perform each part to ensure that an operation cannot be performed by a single entity. Auditing to keep track of information coming in and going out is also a requirement in this model.

2.3.2.4 Brewer and Nash Model (Chinese Wall Model)

Conflict of interest is the key concept in this model. No information is allowed to flow when there is a conflict of interest between the subjects and object [3]. Like the Bell-LaPadula and Biba models, this model also falls in the category of information flow model.

The abstract concepts involved in the model are object, group, and conflict classes. The objects contain information about only one company (e.g., files), the group is a collection of objects of a particular company, and conflict classes are groups of objects that are clustered to compete. The access rule states that an object can be accessed by a subject if the object belongs to the same dataset as that of an object that is already accessed by that subject or belongs to other conflicts of interest class.

This model is based on a combination of free choice and mandatory rules [4]. At first, the subject is free to make a choice and choose an object; once this choice is made, the remaining operations have to be around the dataset of the object.

2.3.2.5 Graham-Denning Model

This model addresses the issues of defining and modifying security and integrity ratings and ways to delegate access rights which were not considered in earlier models.

The model defines what a subject can execute on an object. The model is based on concepts of subject, object, rights, and an access control matrix. The matrix has one row for a subject and one column for a subject and an object. The element of the matrix denotes the rights of a subject on another subject or object. There are eight primitive protection rights that can be issued by subjects on other subjects or objects.

2.3.2.6 Harrison-Ruzzo-Ullman Model

Also known as the HRU model, this is a variation of the Graham-Denning model. This model is based on commands. The commands in turn involve conditions and primitive operations [2].

A difference in this model from the Graham-Denning model is in terms of notation. In HRU, every subject is also an object, and so the columns of the control matrix contain all the subjects and all the objects that are not subjects. The model basically shows how access rights can be changed as well as how the creation and deletion of subjects and objects can be performed.

2.3.3 Cloud Computing Security Models

The cloud computing concept emerged to provide a centralized pool of resources for customer use. It is a concept by which customers are able to avail the hardware, software, and data resources dynamically according to their actual usage. Cloud computing as defined by the National Institute of Standards and Technology (NIST) is widely accepted by researchers. As defined by NIST, cloud computing has five essential features, three service models, and four deployment models. The five essential features include: computing resource pool, virtualized access to a broad network, rapid elasticity, self-service on-demand, and measured service. The three service models are Infrastructure as a Service (IaaS), Platform as a Service (PaaS), and Software as a Service (SaaS). The four deployment models include private cloud, community cloud, public cloud, and hybrid cloud. Here four security models of cloud computing are briefly stated [5].

2.3.3.1 Cloud Multiple-Tenancy Model

This model proposed by NIST is based on the multiple-tenancy function characteristic of cloud computing. Multiple-tenancy [6] allows multiple applications of cloud service providers that are running on a physical server to be available to the customer. This is possible through the process of virtualization. Virtualization is the core technology of cloud computing in which one physical machine can run multiple virtual machines (VM) [7]. The different customer applications are run in different VMs, which enables to isolate fault, virus, and intrusion separately for each service. Thus this reduces the damage of malicious applications. This is a method to satisfy customer demands related to security, segmentation, governance, etc., through implementation by virtualization.

2.3.3.2 Cloud Risk Accumulation Model of CSA

The dependency among the layers in the cloud service model poses risks to cloud computing. PaaS depends on IaaS, and SaaS depends on PaaS, and the security risks of one layer is taken over by the other layer just as the services of one layer is inherited by the other [6]. The IaaS layer implements few security functions and capabilities and hence expects that the customer implements security of operating systems, software applications, etc. The security functions and capability of PaaS are not complete, so customers can implement additional security. In the context of SaaS, more security responsibilities are taken by the cloud service providers than the customers. Hence, the lower the service layer, the more the responsibility of the customer in regard to security.

2.3.3.3 Jerico Forum's Cloud Cube Model

This model represented in Figure 2.3 outlines security attribute information that is present in the service and deployment models of cloud computing. The cloud cube model includes the following parameters [8]:

- Internal/External – Data storage physical location definition. The location of the data storage can be internal (if it is within the boundary of the data owner) or external, but it cannot be concluded that internal data sources are more secure than external ones. A combination could give a more secure environment.
- Proprietary/Open – This parameter is in the context of ownership of the technology, service, and interface. It determines the degree of interoperability. In the case of proprietary, the customer does not have the freedom to change the cloud service, whereas in the case of open more service providers are available and lesser constraint on data sharing is possible.

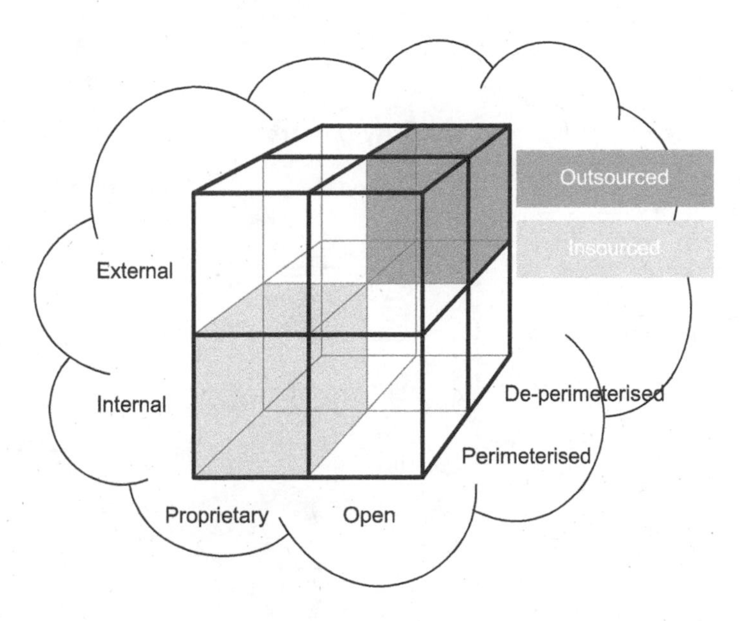

FIGURE 2.3 Jericho Forum's cloud cube model [8].

- Perimeterized/De-perimeterized – Perimeterized refers to the operation of customer's applications within the firewall, but in the case of de-perimeterized the traditional firewall has less function, and a cooperative architectures framework is used to summarize the customer's data.
- Insourced/Outsourced – The parameter defines if the cloud service is presented by the organization's own employees or by a third party. The point to be decided according to the parameter is regarding the management of the cloud services.

2.3.3.4 Mapping Model of Cloud, Security, and Compliance

This mapping model in Figure 2.4 allows analyzing of the gap and identifying the security control strategies that should be provided by cloud service providers, customers, or third parties. The effective protection of the cloud environment is possible when the security risks are identified, and the gaps are recognized according to the architecture and its compliance framework as proposed [6]. The mapping model contributes to the acceptance or refusal of the security risks of cloud computing.

2.4 SECURITY CHALLENGES AND SECURITY REQUIREMENTS IN INTERNET OF THINGS (IoT) ARCHITECTURE

The paradigm shift to connecting the physical and digital world using some communication medium is termed IoT. This has given rise to the Fourth Industrial Revolution, where everything can be connected to the internet. This has led to a boundary-less environment, and systems are facing intrusion and threats from the internet. In the

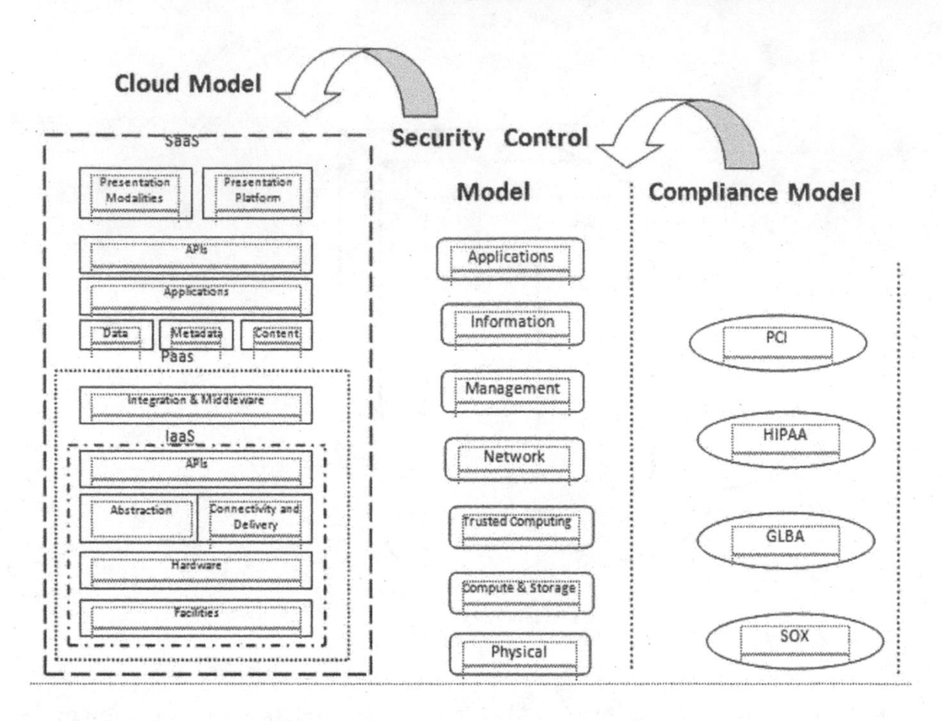

FIGURE 2.4 Mapping model of cloud, security and compliance [6].

IoT-enabled manufacturing sector security challenges are in the area of the supply chain, big data, and industry control systems. Though there are many benefits like reduction of inventory cost, reduction of human interaction, and enhanced monitoring capabilities, possible vulnerabilities demand a proper security system to realize the benefits of IoT. A reference model of IoT infrastructure [9] shows the points of information exchanges that could be vulnerable. IoT systems produce an enormous amount of data, and keeping these data secure, ensuring both protection and privacy, is one of the challenges [10]. The meeting of the internet and the physical world involves threats from manipulation to control. Further, the addition of new devices and protocols expands the attack surface. It is predicted that 28.5 billion devices will be connected to the internet by 2022 [11]. This prediction in a way ascertains the increase in the amount data to be stored and the attacks upon them. The type and nature of attacks are expected to be more sophisticated, and thus security measures with respect to the protection of data from unauthorized users and services form a significant issue in IoT architecture.

Different authors have identified potential threats and attacks to the IoT environment and have categorized the threats and attacks into some general categories. The categories are identified as communications, device/services, users, mobility, and integration of resources [Pal et al. 2020]. Threats in the communication category can be further grouped into routing attacks, active and passive data attacks, and flooding. In the case of devices/services, threats could be physical attacks, device access attacks, device subversion attacks, and device degradation attacks. In users, category threats could be from the point of identity, trust, behavior, and confidentiality. Mobility related threats include tracking and location privacy, dynamic infrastructure, and more than one jurisdiction. As in an IoT environment data collection, data processing, storage, and usage of data require a diverse infrastructure, and there is a security threat at every stage. Thus the threats can be divided into cross-domain administration, interoperability, and cascading resources.

The threats and vulnerabilities of an IoT-based system have given rise to IoT security requirements based on the threats identified. Some specific security requirements specific to IoT are discussed here [12].

- Lightweight solutions – Implementing high-level security measures requires resources. So to implement security in the constrained resource scenario, a balance between the computing power of the constrained IoT devices and the cryptographic techniques employed must be achieved.
- Decentralized Management – IoT systems are huge in scale. Thus security solutions need to be decentralized so that the security provisions are placed nearest to the point of need.
- End-to-End Security – IoT systems are large as well as spread over heterogeneous platforms with multiple domains and technologies. This requires interoperable, secure technologies to ensure identities are verifiable between endpoints. QoS (Quality of Service) parameters need to be protected during transmission as they are the key issues to be secured in electronic communication.
- Identity Management – The identity of the device and the user is an extremely important area where reliable security measures and techniques are essential. Integrating various services seamlessly across different devices and users will require authentication for applications, users, services, and devices. Thus dealing with identity at such a scale in IoT demands innovative security approaches.
- Privacy – The users in the IoT environment will definitely demand privacy when using their services. Thus privacy issues form another security requirement.
- Mobility – The IoT environment allows the users to be mobile and hence has to deal with highly dynamic systems. Thus security solutions must cover the requirements of the users and ensure smooth information sharing between different devices. Mobile access to data and applications should also be supported.
- Scalability – The IoT environment supports heterogeneity, and systems continue to grow. As the system operates, it scales up, and security must be provided for this style of deployment. Security solutions must allow things and users to join and leave the system.
- Robust and reliable – Security solutions for IoT systems should be able to self-repair. They should be able to find faults and rectify them automatically. Real-time addressing of the threats is necessary due to the active nature of IoT.

There are various layers of IoT security architecture. Every IoT security architectural layer has different security requirements. The above requirements form part of the various layers of IoT security architecture. An outline of the security requirements of the layers is as follows:

- Device sensing layer – lightweight solutions, robustness, reliability, mobility, and interoperability.
- Network management layer – interoperability, lightweight solutions, scalability, and mobility.
- Service composition layer – heterogeneity, self-healing, real-time, reliability, and interoperability.
- Application layer – robustness, reliability, interoperability, composition, and privacy.
- User interface layer – mobility, incremental deployment, composition, interoperability, and privacy.

2.5 ADDRESSING SECURITY CHALLENGES IN IoT WITH ARTIFICIAL INTELLIGENCE (AI)

IoT security challenges are unlike the conventional network challenges due to the limited resource availability of IoT devices. IoT devices use LoRa, ZigBee, and other less secure wireless communication media, as complex security protocols cannot be implemented in them due to limited resource availability. Moreover, IoT devices have various data formats that have made it very difficult to develop standard security protocols. Several classes of security threats identified in IoT domains are denial-of-service, man-in-the-middle, and malicious software. In addition to this, IoT users are also exposed to privacy attacks like sniffing, de-anonymization, and inference attacks [13]. Several machine learning (ML) algorithms have been proposed by different researchers, which have proved very helpful in addressing security and privacy issues. ML algorithms are categorized into the following categories:

- Supervised learning – In this case, the target is specified, or we can say it is known for the set of given inputs. The rules are identified in the process of learning to define the classes and then predict the elements.
- Unsupervised learning – Here the data are grouped without the need for human intervention. Unlabeled data clustering and discovering the differences and similarities make it suitable for exploratory data analysis, segmentation, etc.
- Reinforcement learning – The learning, in this case, is reinforced by the feedback of the action. The basic objective of the agent in this learning process is to find that action that maximizes the total reward.

Machine learning algorithms are used to address issues related to authentication, attack detection, intrusion detection, etc., in the IoT domain. Classification and regression are used for modeling the available data sets and predicting continuous numeric variables. Clustering, which is an unsupervised technique, is used for grouping data on the basis of similarities and dissimilarities. Reinforcement learning uses the action and reward relationship, which help in solving many IoT-related problems [14]. A survey of the present machine learning-based solutions for IoT security has listed the following [13]:

Authentication and Access Control in the IoT – The different mechanisms of access control in IoT divides the control mechanisms into different categories. Many ML-based authentications and access control mechanisms have also been proposed. Some of the techniques include a physical layer authentication scheme and using an ANN-based authentication mechanism.

Attack Detection and Mitigation – IoT systems suffer from attacks to their different layers in various degrees, from low scale to high scale. The traditional way of attack detection and mitigation based on cryptography is less accurate. Therefore unsupervised learning, K-NN, and SVM techniques are used to improve the detection of attacks and mitigation in IoT.

Denial-of-Service (DoS) and Distributed DoS (DDoS) Attacks – The IoT domain is referred to as the "Land for DDoS attackers" by Vlajic et al. [15]. As traditional measures, DDoS detection and mitigation means are applied at the entry points of the IoT networks. Researchers have suggested K-NN, decision trees, neural networks, and SVM to detect DDoS in IoT.

Anomaly/Intrusion Detection – Traffic classification and models based on behavior are used for anomaly/intrusion detection in the context of IoT. Zero-day intrusions, which are possible in the case of IoT cannot be detected by the above-mentioned methods. Hence many researchers have proposed AI-based mechanisms and ML techniques like K-means clustering, decision tree, and a hybrid of the two techniques to improve accuracy in detection and reduce false positives.

Malware Analysis in IoT – The IoT domain also faces the challenge of malware in areas of security, authorization, authentication, and also software modification in IoT devices. Spyware, trojan, bot, and ransomware are a few common types of malware. Research works suggest detecting Android-based malware can be done using supervised learning with a random forest classifier [16], using SVM to detect malware and their propagation [17], and also to detect false data injection [18]. Unsupervised learning techniques are also used by authors to isolate the normal data from tampered data to detect attacks.

Present IoT services are mostly cloud-based as all the data processing and analysis is performed in the cloud. The increasing rate of IoT implementation has posed challenges to cloud computing in terms of supporting rapid development. Edge computing is a technology that is recently getting widespread attention as it has many advantages over cloud computing. Edge computing is advantageous in protecting end-user privacy, reducing the burden of network width and energy consumption of the data center. In this edge computing technology, the data generated by IoT devices need not be transmitted and processed in the cloud, thereby reducing data transmission, which is efficient in terms of reducing the time required for data computing. The security issues related to edge computing are due to the distributed layout, limited computing source, and heterogeneous environment [19]. The existing security measures are not adequate enough to combat the continuously upgraded methods of attack. In this scenario, the emergence of AI-based techniques is able to provide new solutions to security issues.

In the area of intrusion detection, machine learning algorithms can be used to extract the data access patterns from training data which can be used to identify intrusions. This can enhance the efficiency of intrusion detection [20, 21]. In the case of privacy preservation, the existing methods like anonymization, cryptography, and data obfuscation require a high number of resources. This makes it difficult to implement edge computing technology as the edge nodes in edge computing have resource constraints. The distributed machine learning method can be applied to reduce the burden on the edge nodes in this case [22]. In an IoT environment, different devices have authorities assigned to perform operations and access the nodes. Thus only authorized access and operations are allowed within the nodes [23, 24]. A classification algorithm of machine learning can be used to classify the privileges of IoT devices [25]. This classification of the devices can be further utilized for access control and preventing attacks.

Though ML can be used to address security issues in IoT, there are certain aspects that need to be addressed in this area. There are some limitations to the ML-based security solutions that require addressing before implementation. ML systems carry risk because of the training data set, which is external to the system. Thus protecting data becomes a challenge. False prediction, data poisoning, data privacy, and confidentiality are some areas along with noisy data challenge that is to be handled for ML systems. ML algorithm performance degradation due to misclassification is another area of concern in ML systems.

2.6 STUDY OF RESEARCH WORKS ON SECURITY ARCHITECTURE AND MODELS

This section discusses works by different authors in two areas. One in designing IoT-related security systems and the other in the area of security architecture and models. In research in the cloud computing domain, security has always been a top issue. A role-based security architecture is proposed in a research work where data security, as well as system efficiency, is improved by assigning reasonable authority to the users requiring different security requirements [26]. In another work on security architecture, a layered trust information security is proposed for risk management at different levels regarding information in an organization [27]. Work on Android was also discovered architecture styles used to perform various analyses to uncover inter-app intent communication-related vulnerabilities [28]. Researchers have proposed security architecture for 5G networks with the inclusion of requirements for virtualization and trust issues among stakeholders. The architecture covers the relevant aspects of 5G networks [29]. Malicious attacks on small and medium-sized businesses have posed challenges to information security. A review of the NIST was conducted to propose an information security framework consisting of four phases: Plan, Do, Check, and Act [30]. The proposed model claims to address the flaws of the previous models designed for small businesses. In the area of AI, the lack of explainability causes weaknesses in the security of machine learning systems. Huawei, with its goal to provide a secure AI application environment, has proposed a three-layer defense system for deploying AI systems [31]. It includes attack mitigation, model security, and architecture security.

IoT allows communication through the internet, which overcomes the challenge of limited range communication. IoT-based alert systems are normally designed to alert the user based on the intruder's motion detection. This motion is sensed by a sensor. A Raspberry Pi triggered by motion-sensing sends an initiation signal to the camera, which captures the intruder's image. The saved images are then sent to the user's mail id. The limitation of this system is the requirement of uninterrupted internet services at both the ends, at the place of the installed system, as well as the user. Another idea with the introduction of Arduino with Raspberry Pi and a GSM module interface was proposed to overcome the previously identified limitation [32]. The proposed system architecture alerts the registered mobile user via an SMS or calls to say to check the email. This does not require the user to be constantly online. In another proposed architecture for home automation and security systems, a door sensor informs the user of the door opening in a house or office through an Android application [33]. The architecture uses microcontrollers and a Raspberry Pi, and communication between the devices by using RF signals. In this proposed work on home security systems, the author proposes a system that notifies the user regarding any unauthorized entry. The system architecture uses Arduino Uno to interface with components, a magnetic Reed sensor for monitoring, a buzzer for alarm, and a Wi-Fi module for communication using the internet [34]. The proposed system is a low cost, low maintenance, and easy to set up system. In another work on home security systems based on IoT using a Raspberry Pi and IoT module, the system shows whether a person is authorized to enter the home [35]. This is done by storing the images of the authorized persons in a database. When an image is encountered by the camera, it compares with the stored images. If matching results are found, the person is identified as authorized. Taryudi et al. proposed a home security and monitoring system based on IoT using Arduino-nano and a controller. The proposed system uses an RFID reader and emails to notify the users. PIR (Passive Infrared), DHT (Digital Humidity Temperature), and LDR (Light Dependent Resistors) sensors to detect intruders, detect humidity and temperature and monitored light conditions are used by the system [36].

All the above works by different authors have developed the domain of security architecture to address issues arising from new technology developments and make the use of new technology more secure.

2.7 CONCLUSION

The architecture of a computer system is responsible for ensuring authorized access to the system and keeping the system in a secure state. A security policy is supposed to deal with these aspects. Security models in turn are built to support the security policies. The different categories of models consist of models where each addresses a certain aspect like confidentiality, integrity, information flow, and being in a secure state in every operation performed. The ultimate objective is to keep the subject–object interaction secured.

The security issues in IoT need to be addressed separately as the requirements are different from other environments. The individual layers of the IoT architecture are exposed to different types of vulnerabilities, and need to be addressed separately. Though there could be some overlapping issues in the layers, each layer needs to be examined to identify the potential security threats. This chapter highlights some work on IoT-based home security systems, which tries to alert the user in case there is unauthorized access to the home. Edge computing is a new computational paradigm that can enrich the field of IoT. There are research works supporting the collaboration of machine learning, deep learning, and other AI techniques in the security of IoT. Machine Learning algorithms can give promising security solutions to the dynamic environment in IoT in place of traditional solutions, which suffer from issues dealing with the active nature of the IoT environment.

AI, with its sub-disciplines of voice recognition, natural language processing, game theory, and computer vision are developing. Security assurances to AI systems are required as they are vulnerable to attacks. These attacks may affect the correctness of the decision making or may steal confidential data used to train AI systems. AI system security covering attack mitigation, model security, and architecture security should be addressed before involving in safety tasks related to personal or public interest.

2.8 FUTURE SCOPE

With respect to the security of software, the concept of secure software development is now taking the lead. Taking security measures in every phase of software development is the aim of the concept. With the different domains of software development, including web application development, cloud computing, IoT environment, implementation of artificial intelligence, mobile computing etc., the security aspect is facing more and more challenges in terms of authenticating users and maintaining the integrity of the systems. Security issues in secure communication is also another major area that is posing threats to system security. In such an environment where digital technology has taken a front seat, maintaining the security of the system, data, and devices has become a priority. More and more research and development can enrich this domain to face the challenges faced.

REFERENCES

1. S. Harris, and F. Maymi, "Security models and architecture," in *CISSP certification all-in one exam guide*, Chapter 5, 7th edition, McGraw-Hill Education, 2016.
2. E.Conrad, S. Misenar, and J. Feldman, "Security architecture and design," in *CISSP study guide*, Chapter 6, 1st Edition, Elsevier Inc, 2010.
3. D. F. C. Brewer, and M. J. Nash, "The Chinese wall security policy," in *Proceedings IEEE symposium on security and privacy*, 1989, pp. 206–214.
4. Franky, *Brewer and Nash security model* [Online], 2019. Available: https://franckybox.com/brewer-and-nash-security-model/, Accessed on April 30, 2021.

5. J. Che, Y. Duan, and T. Zhang et al., "Study on the security models and strategies of cloud computing," in *International conference on power electronics and engineering application*, Published by Elsevier Ltd, 2011.
6. Cloud Security Alliance (CSA), "Security guidance for critical areas of focus in cloud computing(v3.0)," December 2011. Available: http://www.cloudsecurityalliance.org/guidance/csaguide.v3.0.pdf.
7. VMware. Inc., "Understanding full virtualization, para virtualization and hardware assist," VMware, Technical report, 2007.
8. J. Formu, *Cloud cube model: Selecting cloud formations for secure collaboration* [Online], April 2009. Available: http://www.opengroup.org/jericho/ cloud_cube_model_v1.0.pdf.
9. M. J. Koster, *Data models for the internet of things. IoT data models* [Online], September 2012. Available: blogspot.co.uk/2012/09/data-models-for-internet-of-things-5.html.
10. Y. K. Chen, "Challenges and opportunities of internet of things," in *17th Asia and South Pacific design automation conference (ASP-DAC), Sydney, Australia*, January 30–February 2, 2012, pp. 383–388.
11. CISCO, *The Zettabyte Era: Trends and analysis* [Online], November 15, 2019. Available: https://www.cisco.com.
12. S. Pal, M. Hitchens, and T. Rabehaja et al., "Security requirements for the internet of things: A systematic approach," *Sensors*, 2020. Available: 10.3390/s20205897.
13. N. Waheed, X. He, and M. Ikram et al., "Security and privacy in IoT using machine learning and blockchain: threats and countermeasures," *ACM Comput. Surv.*, vol. 53, no. 3, Article 1, p. 35, April 2020. Available: https://doi.org/10.1145/nnnnnnn.nnnnnnn.
14. T. Park, N. Abuzainab, and W. Saad, "Learning how to communicate in the internet of things: finite resources and heterogeneity," *IEEE Acc.*, vol. 4, pp. 7063–7073, November 2016.
15. N. Vlajic, and D. Zhou, "IoT as a land of opportunity for DDOS hackers," *Computer*, vol. 51, pp. 26–34, July 2018.
16. N. An, A. Duff, and G. Naik et al., "Behavioral anomaly detection of malware on home routers," in *12th international conference on malicious and unwanted software (MALWARE)*, pp. 47–54, October 2017.
17. W. Zhou, and B. Yu, "A cloud-assisted malware detection and suppression framework for wireless multimedia system in IoT based on dynamic differential game," *China Communications*, vol. 15, pp. 209–223, February 2018.
18. M. Esmalifalak, N. T. Nguyen, and R. Zheng et al., "Detecting stealthy false data injection using machine learning in smart grid," in *2013 IEEE global communications conference (GLOBECOM)*, pp. 808–813, December 2013.
19. Z. Xu, W.Liu, and J. Huang et al., "Artificial intelligence for securing IoT services in edge computing: a survey," *Security and Communication Networks*, September 2020. Article ID 8872586. Available: https://doi.org/10.1155/2020/8872586.
20. M. A. Amanullah, R. A. A. Habeeb, and F. H. Nasaruddin et al., "Deep learning and big data technologies for IoT security," *Comp. Comm.*, vol. 151, pp. 495–517, 2020.
21. X. Chi, C. Yan, and H. Wang et al., "Amplified locality-sensitive hashing-based recommender systems with privacy protection," *Concurrency and Computation: Practice and Experience*, February 2020. https://doi.org/10.1002/cpe.5681.
22. G. Song, and W.Chai, "Collaborative learning for deep neural networks," in *Proceedings of the advances in neural information processing systems*, pp. 1832–1841, Montreal, Canada, December 2018.
23. H. A. Khattak, M. A. Shah, and S. Khan et al., "Perception layer security in internet of things," *Future Gener. Comput. Syst.*, vol. 100, pp. 144–164, 2019.

24. M. M. Hossain, M. Fotouhi, and R. Hasan, "Towards an analysis of security issues, challenges, and open problems in the internet of things," in *Proceedings of the IEEE world congress on services*, pp. 21–28, IEEE, New York, NY, USA, June 2015.
25. L. Li, T. T. Goh, and D. Jin, "How textual quality of online reviews affect classification performance: a case of deep learning sentiment analysis," *Neural Comput. Appl.*, vol. 32, no. 9, pp. 4387–4415, 2020.
26. L. Tao, J. Shen, and B. Liu et al., "A security architecture research based on roles," in *MATEC web of conferences*, 2017. Available: 10.1051/mateconf/201710005068.
27. R. D. O. Albuquerque, L. J. C. Villalba, and A. L. S. Orozco et al., "A layered trust information security architecture," *Sensors*, vol. 14, pp. 22754–22772, 2014. Available: 10.3390/s141222754; www.mdpi.com/journal/sensors.
28. B. Schmerl, J. Gennari, and A. Sadeghi et al., "Architecture modeling and analysis of security in android systems," in Tekinerdogan B., Zdun U., and Babar A. (eds) *"Software architecture," European conference on software architecture*, 2016, Lecture Notes in Computer Science, vol. 9839. Springer, Cham. Available: https://doi.org/10.1007/978-3-319-48992-6_21.
29. G. Arfaoui, P. Bisson, and R. Blom et al., "A security architecture for 5G networks," *IEEE access*, April 2018. Available: 10.1109/ACCESS.2018.2827419.
30. Y. Alshboul, and K. Streff, "Analysing information security model for small-medium sized businesses," in *21st Americas conference on information systems, Puerto Rico*, 2015.
31. Huawei, "AI security White paper," 2018. Available: www.huawei.com, Accessed on April 23, 2021.
32. B. Aishwarya, S. M. Bindu, and S. V. Molugu et al., "Design and implementation of IoT based intelligent security system," *Int. J. Adv. Res. Sci. Eng.*, vol. 7, no. 7, pp. 290–297, 2018.
33. M. A. Hoque, and C. Davidson, "Design and implementation of an IoT-based smart home security system," *Int. J. Networked Distrib. Comput.*, vol. 7, no. 2, pp. 85–92, April 2019.
34. A. Anitha, "Home security system using internet of things," *IOP Conf. Series: Materials Science and Engineering*, vol. 263, p. 042026, 2017. Available: 10.1088/1757-899X/263/4/042026.
35. Tanaya, K. Vadivukarasi, and S. Krithiga, "Home security system using IoT," *Int. J. Pure Appl. Math.*, vol. 119, no. 15, pp. 1863–1868, 2018.
36. T. Taryudi, D. B. Adriano, and W. A. C. Budi, "IoT-based integrated home security and monitoring system," *International Conference on Electrical, Electronic, Informatics and Vocational Education (ICE-ELINVO)*, September 13, 2018, Yogyakarta Special Province, Republic of Indonesia, Journal of Physics: Conference Series, vol. 1140, no. 012006.

3 The Role of IoT in the Design of a Security System

Keshav Kaushik

CONTENTS

3.1 THE RISE OF CYBER THREATS IN IoT

The Internet of Things (IoT) is expanding at a very fast pace. The advancement in technology of houses, companies, and cars worldwide in relation to the number of connected devices is anticipated to expand from approximately 8 billion now to over 24 billion by the end of the decade, with 5G enabling most of this development. This increase, on the other hand, provides more chances for hackers, many of whom will extend their activities by exploiting weaknesses in sometimes badly maintained IoT devices, to accomplish their objectives. Although fairly inconsequential, linked gadgets offer cybercriminals exposure to enormous quantities of private information stored on their system, rendering them an appealing target. As a result, the protection of IoT devices is critical. In 2017, the outbreak of a malware named NotPetya demonstrated the necessity of securing IoT equipment. Maersk, a global transport and logistics company, was badly impacted, with a total loss of $10 billion [1] to other companies throughout the world.

Scientists claim the exposure of connected gadgets that are linked to the internet is a good indication of IoT infection rates, with significant IoT infection rates happening when gadgets are linked to a publicly accessible IP address. The infection rate is lowered in companies where NAT is implemented on a firewall or router, because IoT devices are not accessible to network scanning. It's essential to note that although conventional NAT converts a private IPv4 domain to a public IPv4 address, carrier-grade NAT adds an extra translation layer as a security safeguard, which will further secure corporate IoT devices. Numerous e-businesses and retail establishments are turning to IoT-based technologies for

DOI: 10.1201/9781003229704-3

their sales, marketing, profitability, and advertising as technology advances. These IoT-based technologies are extremely beneficial to both the owner and the customer. These approaches, however, are subject to a variety of privacy and security problems. The increase of cyber risks in IoT, the corporate view of IoT for e-business and retail security, developments in e-business and retail as a result of IoT, and the flow of security threats linked to privacy and security issues in retail and e-commerce are all discussed in this paper [2]. The Internet of Things is evolving , and it already includes a wide range of technical devices and functionalities. There are several e-solutions available, including e-health, e-transport, home automation, and e-manufacturing, among others. The increasing number of cyber assaults on system frameworks, as well as ecosystem inherent vulnerabilities, is a cause of worry for both manufacturers and customers in this regard. These safety problems must be solved in order to restore consumer confidence and enable the IoT to realize its full potential. This article [3] examines several key IoT services and applications from the viewpoint of hardware, firmware, and software framework configurations, as well as the cybersecurity problems that are anticipated to drive IoT innovation in the immediate future.

The Industry 4.0-recommended regime may exacerbate security weaknesses in contemporary Industrial Internet of Things (IIoT) [4] environments, Industry 4.0 diverse virtual screening pertaining to the physical realm. This is particularly true when such systems receive streams of data from numerous intermediates, demanding multilayer security techniques for connection encryption. Major issues regarding how to safeguard these infrastructures occur when considering the multiplicity of capabilities in the IIoT environment, as well as non-institutionalized hardware and software interoperability. Attacks [5] on power grids may bring entire countries to a stop, resulting in massive economic and financial losses. As a result, security is a key aspect to address before deploying IoT-based smart grid networks on a wide scale. IoT devices [6] transmit data utilizing protocols based on a centralized design, which might lead to data security concerns. Combining machine learning with 5G wireless systems addresses a number of challenges, including autonomous robotics, self-driving automobiles, virtual reality, and security concerns. The system's major aim is to build confidence among network users without relying on a third-party authority.

IoT devices are prone to cyber threats and vulnerabilities; some common vulnerabilities related to IoT devices are mentioned in Figure 3.1. Common vulnerabilities of IoT devices include – malware, exploits, and poor user practices. Malware is malicious software that is programmed to cause damage to and data theft from IoT devices. Malware is of several types; some common malware that targets IoT devices are – ransomware, botnet, worms, and trojans. Such malware is very dangerous in nature and can lead to data theft, data loss, spying, etc. The exploits are used to take control of the IoT devices once the vulnerabilities are identified. The exploits are pieces of code that can gain complete access to the IoT device. Some common types of exploits related to IoT devices include – buffer overflow, SQL injection, network scanning, zero-day exploits, and command injection. The most dangerous among all these are zero-day exploits due to the unavailability of its patch. command injection and SQL injection are also part of the OWASP top 10 IoT vulnerabilities. Poor user

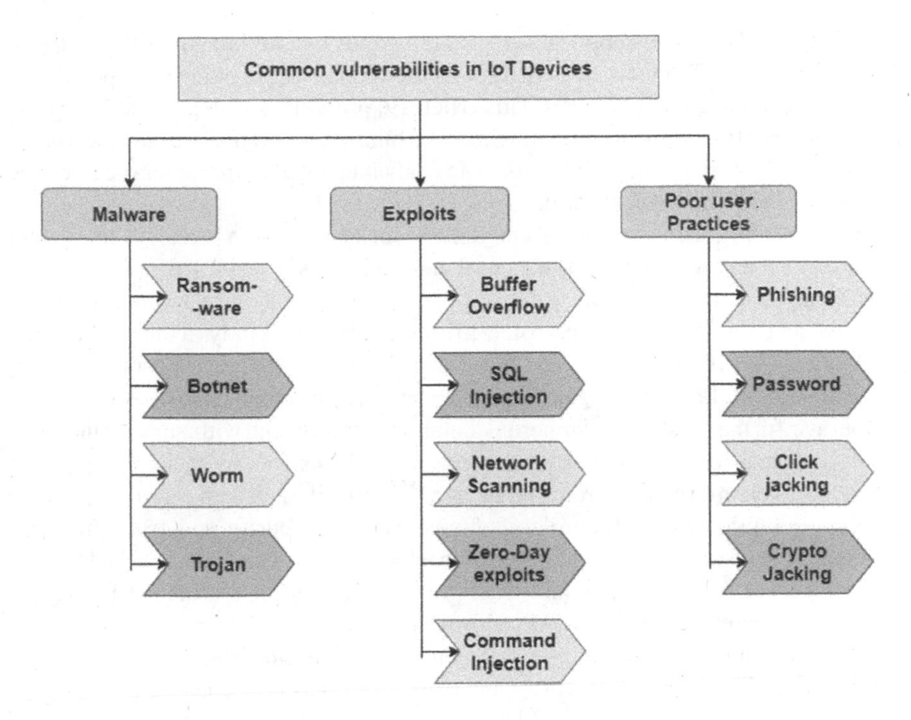

FIGURE 3.1 Common vulnerabilities in IoT devices.

practices may also lead to cyberattacks on IoT devices. Poor user practices include – phishing, poor password, click-jacking, and crypto-jacking (Figure 3.1).

IoT delivers smart solutions to challenges encountered by corporations as well as everyday difficulties experienced by ordinary individuals. Cyber risks [7] are becoming more prevalent as a result of the IoT ecosystem. Cyberattacks primarily target network architecture, but they can also target system vulnerabilities, which are of importance to both manufacturers and customers. Cyberattacks are not new to the IoT; however, as technology becomes more prevalent in our daily lives, it's become important to address this problem and take steps to mitigate the risk. The ever-expanding and diverse network carries a slew of security flaws, dangers, and risks, and is a frequent target of cyberattacks. The establishment of an opportunity and security risk mitigation balance allows people to profit efficiently and effectively from the opportunities available. This study [8] seeks to disseminate the findings of research into security risks, mitigation strategies, and the management balance.

3.2 ENTERPRISE VIEW OF IoT FOR SECURITY SYSTEMS

The Internet of Things offers a lot of potential for updating outdated industrial gear with monitoring capabilities in order to unlock new possibilities and save money. However, implementing the IoT on the production line tier exposes it to new security risks, which may have a substantial detrimental impact on operations. The technique is implemented on widely applicable registered goods by providing real-time adaptation features for IoT device security through component segregation and specialized featherweight integrity protection. This article [9] presents a fresh perspective on IoT endpoint potential problems and mitigation techniques, as well as the advanced area of risk-proof architectures which allow IoT advantages after protecting embedded operations in organizational settings.

By presenting a revolutionary endpoint security-by-design strategy, this paper addresses such essential new threats at their source. As a result [10], many people mistakenly assume that e-IoT systems are safe due to a lack of awareness of the hazards and the expensive cost of e-IoT devices. This study focuses on e-IoT components, e-IoT weaknesses, countermeasures, and related security consequences in order to fill knowledge gaps, increase awareness about e-IoT security, and encourage further research. Numerous computer equipment with surveillance and networking capabilities may now be connected thanks to new technology. These devices gather information from the device, analyze it, and then retain or transfer it to other equipment with similar features. Several security and trust problems develop as a result of this communication mechanism, which must be addressed immediately. Certain security and trust problems might pose a danger to the IoT ecosystem; therefore, they must be addressed using the most up-to-date strategy, methodology, and innovation. In [11], the authors aim to illustrate some of the most significant challenges to IoT security and trust, as well as how contemporary design addresses them.

Businesses may get a competitive edge in today's increasingly competitive industries by successfully implementing enterprise resource planning (ERP) solutions. ERP integrates a variety of technologies, such as the Internet of Things. To recognize, regulate, and send data to persons and databases, the IoT employs a distinct set of rules. Data is gathered via the IoT, saved in the cloud, and then retrieved and managed using ERP. ERP [12] can monitor, regulate, and analyze the data because it is kept in the cloud. Consequently, integrating ERP with the IoT opens up a slew of possibilities, the most notable of which are improved management, robotics, and quality control, as well as lower ERP operation cost. By decreasing human interaction and boosting automation through sensors, the Internet of Things establishes a link between the product and the client. However, integrating the IoT into ERP is fraught with difficulties, the most significant of which is a lack of faith in IoT as a comparatively recent technological revolution.

Enterprise information systems (EISs) support all data gathering, business intelligence, connectivity, and associated decision-making processes. As a result, the network infrastructure for data gathering and exchange has a significant impact on the performance of an EIS. The goal of this study [13] is to look into the influence of security on enterprise resource planning in the IoT paradigm. The IoT's groundbreaking potential opens up a plethora of new marketing strategies for providing quality across sectors, goods, and service offerings. Ensuring IoT technology is dependable and safe, on the other hand, is critical to achieving the full potential of this game-changing notion. This paper [14] outlines a method for detecting IoT loopholes in businesses. The IoT is becoming imperative in businesses.

IoT safety differs from typical PC information security due to the size of the IoT, and convenience and uniformity of such smart devices, and the ability to identify assaults using sensor data. These characteristics of the Internet of Things, along with the huge computational power of upcoming hardware devices, may be leveraged to create a security assessment tool tailored to IoT security. Because ERPs are linked to the internet and intranet, ERP security has become a problem for major businesses. The problem [15] has worsened since the growth of IoT that allows businesses to control different elements of their operations by connecting several devices to a network. Switching to linked IP systems not only enables automation, but it also reduces the cost of operation and complexity of existing systems. ERP systems are copyrighted software designed to be utilized within the four walls of an organization, making them more vulnerable to hackers.

EIS management systems should accurately evaluate the massive data created by IoT devices connected in India's smart cities and utilized by businesses. The emergence of less costly digitalization and cognitive computing, as well as pervasive flexibility in the architecture of information technology concerns, has resulted in an inflow of services. As a result, business functions have gotten simpler. However, the accessibility of this fully engaging and simple program would necessitate adequate management of privacy and security issues about the data gathered. This work [16] has made a comprehensive effort to evaluate and identify variables relating to enterprise-related concerns, particularly ethical ones, that might finally offer adequate security to IoT-enabled devices utilized in India's smart cities.

3.3 FUTURE OF IoT IN CYBERSPACE

The IoT is a collection of physical items that can recognize themselves to other gadgets and utilize integrated technology to communicate with intrinsic or extrinsic states. The IoT is a setting that can describe itself and grows in importance by communicating to other items and the vast amounts of data that flow around it. Increasingly personal and commercial data will be stored in the cloud and sent across millions of nodes, some of which may have vulnerable flaws. One weak link in the security program may give hackers access to an almost infinite number of doors that might be unlocked, giving them access to confidential information. Cybercriminals may now create attacks with unprecedented complexity and associate data from a variety of private foundations, including vehicles, cellphones, smart thermostats, and even freezers, in addition to public connections. As the sector reinvents and creates products that are networked with the internet, it must benefit from its errors. Several security best practices, such as strengthening networks, adopting secure communication protocols, and applying the most recent updates, corrections, and patches, may be used. Attackers appear to be everywhere, and their increased focus on the IoT is a logical transition given that they are interested in where the globe's data is moving. The linked society is on its way, but so are the cybercriminals. The bottom line is that the information security environment is already adapting to these new network needs. The harsh reality is that we are still a long way from a utopia in which the Internet of Things controls security seamlessly through networked devices and offers a secure infrastructure for customers and their personal data.

Eventually, we are moving into an era when innovation may be used to generate limitless outputs that can assist in leading a sustainable existence while also opening up new doors to the wider world. This would have a significant influence on health and will make the situation more intelligent and controllable. New technologies, which were previously a means of progress, are now being transformed into biochemistry. Organizations have built web applications that respond to typical shopper demands as a result of this anytime/anywhere internet connectivity. These programs may have everything from tracking meal portions to communicating large amounts of data with a single mouse click. We are on the verge of an IoT transformation as companies and consumers speed up adoption. In [17], the authors went through security issues that have been observed in IoT devices, emphasizing how susceptible these gadgets are when confronted with current, advanced cyberattacks. The writers also examine how to address the security issues raised by IoT. The functioning of underpinning wireless paradigms like WSNs, SDR, CR, and RFID is heavily responsible for large, efficient connectivity among different IoT applications. Researchers provide topical layered classifications to emphasize potential vulnerabilities, threats, and problems of major IoT enabling technologies that support smart city growth. The authors [18] also highlight important facilitators and possible needs that are critical to the creation of sustainable smart cities. Furthermore, the authors went through some of the open issues that need to be resolved in order to fully realize 5G's prospects for the development of smart cities.

Researchers explain the cyber-enabled Internet of X (IoX) [19] from the perspectives of both omnipresent interconnections and spatial confluence, and propose a framework with four pillars, particularly objects, humans, cognition, and cyber entities in their respective environments. Various device makers use various network systems on their devices. The application layer, transport layer, and network layer are three network levels in which we investigate the characteristic areas of IoT devices. Using network protocol features, we create IoT device fingerprinting using neural network techniques. Researchers [20] also put the prototype system into action and do real-world tests to see how well device fingerprints work.

Given that smart networks account for the majority of cyberspace's traffic, as well as the numerous gadgets that make up the IoT, a happy relationship between IoT and 5G seemed unavoidable. This consequence [21] is a new paradigm of the internet in which little gadgets such as coffee makers communicate with cellphones in the same way that the home fridge communicates with the garage door opener. It is a political, moral, and security conundrum in the untamed west. All of this chaos is the consequence of our complete reliance on technology, which is increasingly infiltrating our homes and bedrooms.

3.4 FUNDAMENTALS OF SECURITY SYSTEMS

All home security systems are built around the same basic concept of protecting entry points such as windows and doors, as well as inside storage space for valuables such as art, computers, weapons, and cash. The only major difference, regardless of the value of your home, the number of windows and doors, or the number of interior rooms a homeowner chooses to protect, is the set of security mechanisms scattered throughout the residence and controlled by the control panel. The name of any security system will provide you with the most basic explanation. It is, in essence, a method or methodology for safeguarding anything via a collection of interrelated devices and systems. The following are the components of a typical home automation system:

- The principal operator of a home's surveillance system is a control panel.
- Sensors for doors and windows.
- Interior and outdoor motion detectors.
- Surveillance cameras can be wired or wireless.
- A buzzer or alert with a high decibel level.
- Windshield decals and a yard sign.

House security systems [22] are built on the basic concept of employing sensors to guard entry points into a house, which communicate with internet gateways or command centers situated in a convenient location within the building. The sensors are generally installed in front and back doors, as well as readily available windows, specifically those that open, particularly those at ground level. Motion detectors can be used to safeguard open spaces within houses.

- The control center is a microprocessor that activates and executes surveillance systems, interacts with each piece of installed equipment, generates an alarm unless demilitarized zones are breached, and communicates with an alert monitoring company.
- Sensors for doors and windows are made up of two pieces that are put next to each other. The gadget has two parts: one that goes on the window or door and the other that goes on the entrance frame or windowsill. The two components of the sensor are linked together when a door or window is closed, forming a security circuit.
- If motion detectors are turned on, they may have a blind spot that can be infiltrated without causing an alarm. These are frequently used to protect valuables in rooms and less-frequented areas of larger homes.
- Security cameras may be used in a variety of ways as part of a larger security system, and are provided in both wired and wireless variants.
- Home security alarms, which are loud enough to be heard by neighbors, serve a variety of functions. First, they notify the occupants of the residence that a problem has arisen.

A smart controller connects to a Wi-Fi network, allowing you to track and control your security equipment using your smartphone and an app. The locking system usually includes a gateway that links to these devices by one or more communication technologies such as Wi-Fi, Z-Wave, Zigbee, or bespoke network segments. You may add locks on the doors, key fobs, interior and outdoor video surveillance, lights, alarms, burn detectors, moisture sensors, and other features to your system. Any effective smart surveillance system will include parts that work together in a seamless environment and can be controlled by rules. For instance, you can program lights to turn on when movement is detected, doorways to open when a smoke detector sounds, and a webcam to start recording when a sensor is activated. All [23] of the devices we evaluated includes an app that allows you to disarm and disable the system, establish rules, add and delete devices, and get push notifications when alerts are activated using your smartphone as your command center.

3.5 SECURITY ASSESSMENT OF IoT SOLUTIONS

By integrating items to the internet, the IoT paradigm may be used in smart homes, posing additional privacy and security concerns from the perspective of the privacy, legitimacy, and validity of information sensed, gathered, and transferred by IoT gadgets. Because of these issues, smart houses are very susceptible to many sorts of security assaults, making IoT-based smart homes unsafe. The study [24] results main objectives are to showcase the different security issues of IoT-enabled smart homes, to communicate the threats to house residents, and to suggest ways to mitigate the discovered dangers. The use of the analytical hierarchy process (AHP) [25], a popular approach for a multi-criteria analysis, for IoT risk evaluation is presented. Prior to establishing preparedness and prevention plans for IoT information security, the risks are prioritized and graded at distinct tiers, with a well-defined IoT risk vocabulary including 25 hazards across six layers of the IoT architecture. The individuals and procedures layer, the applications layer, and the top three important layers include the network layer, with threats such as lack of awareness, virus infiltration, malicious software implant, denial of service, and insufficient IoT regulation having the highest priority and rating.

Several businesses are hesitant to use such solutions because of security concerns, while others just neglect security concerns while incorporating CloudIoT [26] into their operations. As a result, with so many cloud services and IoT devices to choose from, determining their secure communication becomes a critical problem for promoting CloudIoT adoption and reducing potential company risks. We provide an end-to-end vulnerability assessment technique relying on a software-defined network (SDN) to establish the threat level for a specific CloudIoT offering in order to address this problem, taking into consideration the importance of corporate data in CloudIoT. If cybersecurity cannot be quantified, it cannot be prepared for, managed, analyzed, or regulated. Nevertheless, inexperienced IoMT users face challenges when it comes to selecting security solutions that are both adequate and resilient. As a result [27], we created the IoMT Security Assessment Framework (IoMT-SAF), a web-based IoMT security evaluation technique for suggesting safety features in IoMT and assessing defense and deterrence in IoMT solutions based on a unique metaphysical set of circumstances approach. IoMT-SAF aids in the discovery of a solution that satisfies the customer's strategic goals as well as the decision-making process.

Numerous initiatives have been launched to address the increasing security problems in IoT systems and to make them self-sufficient in terms of energy harvesting for smooth operation. In light of these facts, the authors [28] investigate the growing vulnerabilities in IoT devices in this article. We include a cutting-edge survey that covers many aspects of the Internet of Things. The article also includes a thorough taxonomy of exploits linked to various flaws. Then we concentrated on IoT vulnerability assessment methods, accompanied by a case study on smart agriculture's long-term viability. IoT applications are particularly sensitive to many forms of security threats as a result of the challenges, culminating in unsafe IoT environments. As a result, conducting a scenario assessment to detect potential security threats is required in order to build a full picture of safe IoT installations. The essential parts of IoT models are presented in this article, as well as a scenario evaluation for IoT applications. Characterized by three domains of the IoT paradigm, the authors [29] have identified security enhancement techniques for IoT applications. Developments in the security arrangements evaluation will be emphasized, as well as the determination of specialty areas where concepts proposed will be focused on.

While evaluating the privacy and security of solutions, consumers are frequently left in the dark. Using the Analytic Hierarchy Process [30], this article provides a methodology for statistically assessing and comparing the privacy and security of Internet of Things technologies. It also includes detailed privacy and security evaluation standards for improving privacy and security in IoT systems. Through encouraging openness and raising security awareness, this project attempts to bridge the gap between customers and suppliers. The P-SCAN test [31] platform is meant to democratize linked device security evaluation, ranging from traditional industrial equipment to IoT goods. The platform includes a collection of test suites that automate the process of evaluating security mechanisms on the device's communication protocols, which is linked to recommendations that make defining a device security target easier. The technology is meant to be extensible and adaptable in order to discover possible weaknesses as technology advances. This document describes the determined business requirements and market niche, as well as the associated business model and a summary of the technological solution. Figure 3.2 shows the security assessment for IoT devices.

The ability to connect, interact, and administer many networked devices remotely through the web is now a possibility. As the number of connected devices grows, hackers will have more possibilities to collect and alter data. IoT security assessments thoroughly examine IoT communities in order to ensure the efficacy of safety protocols, including IoT analysis and design assessment, Device detection techniques, binary exploitation, and firmware code-breaking, and more. Security analysis of linked interfaces, security analysis of IoT platforms, and security assessment of IoT mobile and cloud applications. The goal of the commitment is to improve IoT deployment, procedures, and controls while also improving defensive capabilities. IoT security assessments [32] are carried out in stages, with situations being created for each networked infrastructure asset and comprehensive checks being performed on each element under examination.

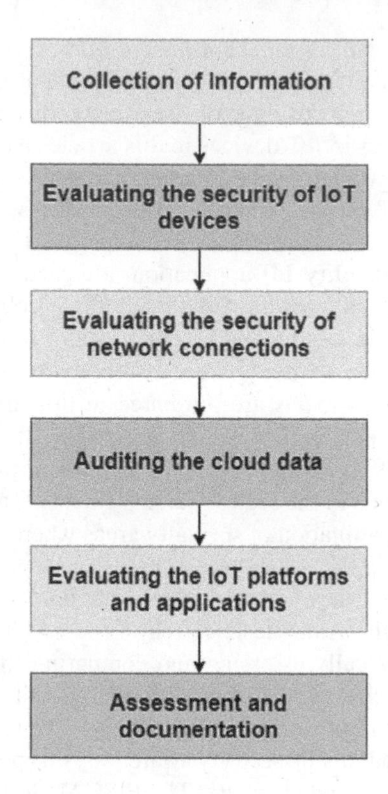

FIGURE 3.2 Security assessment of IoT devices.

- Secure loading should be implemented – Encrypt data at rest or in transit between IoT devices and back-end systems using standard cryptographic techniques to safeguard data security and confidentiality against eavesdropping by hostile actors.
- Set up a robust encryption system – Incorporate access control methods into operating systems to restrict authorized personnel's access to device elements.
- Securely connect depending on roles – By eliminating weak or default credentials, provide role-based connectivity for endpoints.
- Default passwords should be changed – To resolve security flaws, update IoT devices with the most recent firmware and fixes.
- Maintain your gadgets up to date – Using a device id or a device-specific identification, authenticate linked devices.
- Make sure your firewall/IPS is configured correctly and your network is secure – To restrict traffic meant for certain network nodes, set up a device firewall or deep packet inspection capabilities. When not in use, keep an eye on inbound ports and make sure they're closed.
- Continuously monitor device status across IoT networks.
- Deployment of IoT devices in a secure way – While linking IoT devices to residential or industrial circuits, inspect and verify them.

REFERENCES

1. "Defeating the rise of cyber threats with latest TPM 2.0 specification - embedded computing design," https://www.embeddedcomputing.com/technology/security/software-security/defeating-the-rise-of-cyber-threats-with-latest-tpm-2-0-specification (accessed July 30, 2021).
2. K. Kaushik, and S. Dahiya, "Security and privacy in IoT based e-business and retail," *Proc. 2018 Int. Conf. Syst. Model. Adv. Res. Trends, SMART 2018*, pp. 78–81, November 2018, doi: 10.1109/SYSMART.2018.8746961.
3. S. Tweneboah-Koduah, K. E. Skouby, and R. Tadayoni, "Cyber security threats to IoT applications and service domains," *Wirel. Pers. Commun.*, vol. 95, no. 1, pp. 169–185, May 2017, doi: 10.1007/S11277-017-4434-6.
4. K. Tsiknas, D. Taketzis, K. Demertzis, and C. Skianis, "Cyber threats to industrial IoT: a survey on attacks and countermeasures," *IoT 2021.*, vol. 2, no. 1, pp. 163–186, March 2021, doi: 10.3390/IOT2010009.
5. K. Kimani, V. Oduol, and K. Langat, "Cyber security challenges for IoT-based smart grid networks," *Int. J. Crit. Infrastruct. Prot.*, vol. 25, pp. 36–49, June 2019, doi: 10.1016/J.IJCIP.2019.01.001.
6. A. D. Dwivedi, R. Singh, K. Kaushik, R. R. Mukkamala, and W. S. Alnumay, "Blockchain and artificial intelligence for 5G-enabled internet of things: Challenges, opportunities, and solutions," *Trans. Emerg. Telecommun. Technol.*, p. e4329, July 2021, doi: 10.1002/ETT.4329.
7. A. Dhatrak et al., "Cyber security threats and vulnerabilities in IOT," *Int. Res. J. Eng. Technol.*, 2020, www.irjet.net (accessed April 29, 2022) [Online].
8. E. Sahinaslan, "On the internet of things: security, threat and control," *AIP Conf. Proc.*, vol. 2086, no. 1, p. 030035, April 2019, doi: 10.1063/1.5095120.
9. S. Tedeschi, C. Emmanouilidis, J. Mehnen, and R. Roy, "A design approach to IoT endpoint security for production machinery monitoring," *Sensors*, vol. 19, no. 10, p. 2355, May 2019, doi: 10.3390/S19102355.
10. L. P. Rondon, L. Babun, A. Aris, K. Akkaya, and A. S. Uluagac, "Survey on enterprise internet-of-things systems (E-IoT): a security perspective," [Online]. February 2021. https://arxiv.org/abs/2102.10695v1 (accessed July 30, 2021).
11. K. Kaushik, and K. Singh, "Security and trust in IoT communications: role and impact," *Adv. Intell. Syst. Comput.*, vol. 989, pp. 791–798, 2020. doi: 10.1007/978-981-13-8618-3_81.
12. M. Tavana, V. Hajipour, and S. Oveisi, "IoT-based enterprise resource planning: challenges, open issues, applications, architecture, and future research directions," *Internet of Things*, vol. 11, p. 100262, September 2020, doi: 10.1016/J.IOT.2020.100262.
13. C. Thangavel, and P. Sudhaman, "Security challenges in the IoT paradigm for enterprise information systems," pp. 3–17, 2017, doi: 10.1007/978-3-319-70102-8_1.
14. M. Hewlett Packard Labs, and M. Mowbray, "Detecting security attacks on the enterprise internet of things: an overview," *Hewlett Packard Enterp. Dev. LP*, Jan. 2017.
15. S. Kaushik, "Securing ERP cyber systems by preventing holistic industrial intrusion," pp. 97–112, 2021, doi: 10.1007/978-3-030-69174-5_6.
16. S. Chatterjee, A. K. Kar, and S. Z. Mustafa, "Securing IoT devices in smart cities of India: from ethical and enterprise information system management perspective," vol. 15, no. 4, pp. 585–615, 2019, doi: 10.1080/17517575.2019.1654617.
17. U. N. Dulhare, and S. Rasool, "IoT evolution and security challenges in cyber space," pp. 99–127, January 2019, doi: 10.4018/978-1-5225-8241-0.CH005.
18. A. Akhunzada, S. ul Islam, and S. Zeadally, "Securing cyberspace of future smart cities with 5G technologies," *IEEE Netw.*, 2020, doi: 10.1109/MNET.001.1900559.

19. H. Ning, F. Shi, S. Cui, and M. Daneshmand, "From IoT to future cyber-enabled internet of x and its fundamental issues," *IEEE Internet Things J.*, vol. 8, no. 7, pp. 6077–6088, April 2021, doi: 10.1109/JIOT.2020.3033547.
20. K. Yang, Q. Li, and L. Sun, "Towards automatic fingerprinting of IoT devices in the cyberspace," *Comput. Netw.*, vol. 148, pp. 318–327, January 2019, doi: 10.1016/J.COMNET.2018.11.013.
21. J. M. Kizza, "Evolving cyberspace: the marriage of 5G and the internet of things (IoT) technologies," pp. 259–280, 2019, doi: 10.1007/978-3-030-03937-0_12.
22. "What is a security system and how does it work?" https://www.safewise.com/home-security-faq/how-do-security-systems-work/ (accessed August 1, 2021).
23. "The best smart home security systems for 2021," https://in.pcmag.com/surveillance-cameras/99815/the-best-smart-home-security-systems-for-2020 (accessed August 1, 2021).
24. B. Ali, and A. I. Awad, "Cyber and physical security vulnerability assessment for IoT-based smart homes," *Sensors*, vol. 18, no. 3, p. 817, March 2018, doi: 10.3390/S18030817.
25. S. V. Bharathi, "Forewarned is forearmed: assessment of IoT information security risks using analytic hierarchy process," *Benchmarking An Int. J.*, vol. 26, no. 8, pp. 2443–2467, September 2019, doi: 10.1108/BIJ-08-2018-0264.
26. Z. Han, X. Li, K. Huang, and Z. Feng, "A software defined network-based security assessment framework for cloud IoT," *IEEE Internet Things J.*, vol. 5, no. 3, pp. 1424–1434, June 2018, doi: 10.1109/JIOT.2018.2801944.
27. F. Alsubaei, A. Abuhussein, V. Shandilya, and S. Shiva, "IoMT-SAF: internet of medical things security assessment framework," *Internet of Things*, vol. 8, p. 100123, December 2019, doi: 10.1016/J.IOT.2019.100123.
28. P. Anand, Y. Singh, A. Selwal, M. Alazab, S. Tanwar, and N. Kumar, "IoT vulnerability assessment for sustainable computing: threats, current solutions, and open challenges," *IEEE Acc.*, vol. 8, pp. 168825–168853, 2020, doi: 10.1109/ACCESS.2020.3022842.
29. P. K. Chouhan, S. McClean, and M. Shackleton, "Situation assessment to secure IoT applications," *2018 5th Int. Conf. Internet Things Syst. Manag. Secur. IoTSMS*, pp. 70–77, November 2018, doi: 10.1109/IOTSMS.2018.8554802.
30. F. Alsubaei, A. Abuhussein, and S. Shiva, "Quantifying security and privacy in internet of things solutions," *IEEE/IFIP Netw. Oper. Manag. Symp. Cogn. Manag. a Cyber World, NOMS*, pp. 1–6, July 2018, doi: 10.1109/NOMS.2018.8406318.
31. T. Maurin, L. F. Ducreux, G. Caraiman, and P. Sissoko, "IoT security assessment through the interfaces P-SCAN test bench platform," *Proc. 2018 Des. Autom. Test Eur. Conf. Exhib. DATE*, vol. 2018, no. January, pp. 1007–1008, April 2018, doi: 10.23919/DATE.2018.8342159.
32. "IoT security assessment | Internet of things security testing services," https://www.aujas.com/internet-of-things-security-assessment (accessed July 31, 2021).

4 IoT-Tangle Enhanced Security Systems

Falak Bharadwaj, Shyam Raj, and Arti Saxena

CONTENTS

4.1 INTRODUCTION

The first security systems go way back to the beginning of humanity. Vedic sculptures and ancient stories tell us about the different mechanisms used by the people of that time to protect themselves from different breaches and betrayals. With the evolution of mankind, we became more advanced, and the mechanisms we used also evolved. From using

DOI: 10.1201/9781003229704-4

people as a resource to spy on enemies and potential threats, the use of new technologies, like cameras and audio devices, took off. A security system not only helps to prevent man-made threats but also to cope with natural disasters, e.g., it can be used to find someone stuck under the rubble after an earthquake. Drones are used effectively during floods and tsunamis. Different types of systems can be used for the purpose of security, depending upon the challenges faced. With time, the challenges have evolved and changed as per the demand of the time and the evolution of mankind, which also involves the inventions of humans. The research field has gone over the top to invent and report different types of technologies such as telecom and internet services; things are available not only on the doorstep but at the table itself. The expansion of the internet and networking does bring a sort of independence and transparency among the people and the services they are provided with, but it also makes us more vulnerable to different types of threats that sometimes involve the private space around us. Since the COVID-19 pandemic of 2020, people have started working and doing almost everything from home. This is only possible due to advancement in the technologies researchers and enthusiasts provide us with. This lifestyle is not easy but accessible, and the accessibility come with a risk. According to information recorded by the Indian Computer Emergency Response Team (CERT-In), "3,94,499 and 11,58,208 cyber-security events were detected in 2019 and 2020, respectively". That's exactly thrice the number of attacks reported in the year 2019. More accessibility has left people exposed as security systems are exploited on a daily basis. How many unreported attacks occur would be impossible to estimate. Everyone is affected by these issues, as devices we use in our daily lives are among the machines under threat. It has never been easier to circulate information; however, this also means that inaccurate knowledge can circulate easily. The maintenance of this accessible lifestyle, certainly a dangerous lifestyle, is provided using a branch of technology known as Artificial Intelligence (AI). Artificial Intelligence stands for an area of computer science that gives machines the ability to seem like they have human intelligence. Moreover, Artificial Intelligence enables the designing of machines that have the ability to think. AI has evolved extensively over the past ten years showing enormous and unimaginable inventions that do not involve human intervention after they are completed. It provides independence for the "machine" to work on its own. Hence, it has become a bane and a boon to society. Technology can be used to evolve mankind and destroy it with the same input, and AI is a great example of this. The subfield of AI includes machine learning which is the most basic approach to real-world problems. The services provided through machine learning techniques are certain and field-specific. They have the power to learn from a certain type of data, but they cannot learn everything. The technology that involves devices with the capabilities of knowledge, power, and design to work on their own, learn on their own, and to choose what to do on their own, is categorized under the IoT, which stands for the Internet of Things. As the name suggests, the Internet of Things is a system of interconnected computational devices to make their usage easier and more reliable. The device is integrated with computational devices such as sensors, receptors, cameras, mics, microprocessors, and integrated circuits. It may involve more devices to make it more user-friendly. A great example of an IoT device is Amazon Echo. The device is equipped with a mic and sensors with different controls as assigned by the user, which can be used to operate different mechanical functions upon command, i.e., voice input. If integrated, the user can turn on the fan, turn off the light, play music, and make a coffee. The main issue faced by such kinds of devices is that the data being generated or captured by the device is kept in the database of the device's manufacturer, which is not safe as per the current trends of data breaches. Such challenges with new world problems such as terrorism and cybercrime need advancements in the current security systems. Cryptography can be seen as a reliable solution to this, a technology that seen a lot of buzz around it in the past year. Cryptography is a method of protecting information and communications through the use of codes so that only those for whom the information is intended can read and process it. Moreover, Blockchain is considered a great example of cryptography. Existing Blockchains such as Bitcoin, Ethereum, etc., are examples of cryptocurrency. It faces scalability issues due to a vast number of transactions and users. Proof-Of-Work (PoW) computational requirements are huge and consume more energy increasing power shortage and pollution. Conducting proof-of-work using renewable energy is under consideration. Proof-of-stake is another replacement for proof-of-work. Centralization by powerful miners is a serious problem as the power of Blockchain is in its decentralization. The cost of transactions is present. All integrity guarantees are probabilistic, and privacy requires a bit more thought. Replace Blockchain with Tangle, a type of Blockchain that doesn't have a chain. It borrows a lot of ideas from Blockchain, but it is not exactly a Blockchain, as there is no block – individual transactions are tangled together. What is Tangling? Construct directed acyclic graphs (DAGs) connecting transactions, self-regulating, very scalable. It still uses PoW – but a long overhead PoW that prevents spamming. One of the important features of Tangle is CAP (consistency, availability, and partition tolerance). Other features include that there are no fees, it is easily scalable, modular, lightweight, and it allows offline quantum proof. The research is directed toward IoT systems involving the support of Tangle that can easily strengthen the systems with powerful hash functions and consensus mechanism.

4.2 SECURITY SYSTEMS

As the term suggests, a security system is a system related to security that secures a particular entity. The simplest example of such a system is home security systems, i.e. security systems used to secure a home and stay active until something turns it off. CCTV cameras are the prime example of one of the security systems around us. In public places, security systems appear in different forms, such as metal detectors, sensors, advanced CCTV cameras, fire alarms, etc. Lockers used in houses and banks are also part of security systems.

The use of security systems is more technology-based nowadays, but the concept has evolved throughout history in different forms. In ancient times, kings used to have underground temple stores where they stored their precious items. In order to keep the underground stores safe from invaders, they used to spread fake news and myths about the place of their precious items. As for the other systems used man powers, as we term them, soldiers, to guard the palace and keep an eye on the invaders. They used to have informants in other kingdoms and on their own to keep a check on the news and rumors going around the kingdom. As mankind evolved, humans became more technologically enhanced; the researchers and scientists developed smart and efficient security systems that work electronically and provide better security with more accuracy and precision. A great perk of this enhancement is that humans don't have to move from place to place or stay in a particular place to be updated regarding security. For example, CCTV cameras can be accessed remotely. Different types of security systems are discussed below.

Alarm systems are one of the technical tools available to help with company continuity. Integrating alarm systems is a new method of utilizing the current technical capacity parts of intruder alarm systems, CCTV, access control, and hold-up alarm systems [22]. These apps can be linked with one another or used to augment non-alarm systems, simplifying the automation processes in commercial and residential facilities. This article addresses the issue of technological solutions for interconnecting alarm and non-alarm systems. The article's main contribution is a categorization of approaches for integrating various systems.

Because of the rising dangers and government restrictions, cyber-security solutions that defend networks and computers from cyber assaults are becoming more widespread [18–20]. At the same time, the massive volume of data collected by cyber-security systems poses a severe danger to the privacy of individuals who rely on them. To contextualize this danger, we examine current and innovative cyber-security systems and assess their potential for privacy violations. The article proposes a taxonomy for assessing privacy concerns in information security systems based on the amount of data exposure, the level of identification of individual users, data sensitivity, and user control over monitoring, data collecting, and analysis. It discusses the findings in light of recent technology advances and suggests many new avenues for making these processes more privacy-conscious.

4.3 ARTIFICIAL INTELLIGENCE

The term AI was coined in 1955 by John McCarthy, also known as one of the "Founding Fathers" of AI, along with Marvin Minsky, Allen Newell, and Herbert A. Simon. The famous Dartmouth Summer Conference in Summer 1956 initiated the field of AI. Artificial Intelligence is an area of computer science that gives machines the ability to seem like they have human intelligence, i.e., providing independence to the machine to work on their own. It is designing machines that have the ability to think. It is the intelligence of machines. The concept of Artificial Intelligence is derived from "human intelligence." Human intelligence is the ability to derive information, learn from experience, adapt to the environment, understand, and correctly utilize thoughts and reactions. Mental quality consists of the abilities to learn from experience, adapt to new situations, understand and handle abstract concepts, and use knowledge to manipulate conditions. A great example of human intelligence is talking to a new person or moving to a new house. Human intelligence plays a major part in human behavior as it defines the movement and the lifestyle of human beings depending upon their present surrounding conditions. The computer science field tends to build an intelligence similar to human intelligence to optimize the output and time complexity for major tasks requiring high human intervention. The goal is to allow the machine to think independently and achieve the goal state in optimum circumstances. Some of the major projects powered by Artificial Intelligence in different fields are:

4.3.1 Robotics

1) Hanson Robotics [11] – Building humanoid robots: the Hanson team, founded by David Hanson, Ph.D., has developed a worldwide reputation for producing robots that appear and act truly alive, including the well-known robot character Sophia the Robot. Robots come to life as fascinating characters, practical goods, and developing AI thanks to our advances in AI research and development, robotics engineering, experiential design, narrative, and material science.
2) iRobot [15] – Smarter home robots: the iRobot Corporation is a technology firm based in the United States that creates and manufactures consumer robots. It was created in 1990 by three members of MIT's Artificial Intelligence Lab who were working on space exploration and military defense robots. The firm, based in Delaware, produces a variety of autonomous household vacuum cleaners (Roomba), floor moppers (Braava), and other cleaning equipment.

4.3.2 Health Care

1) Atomwise [4] – Streamlining drug discovery: Atomwise uses AI to revolutionize drug discovery. They pioneered the application of deep learning for structure-based drug discovery, and they now have a pipeline of small-molecule therapeutic candidates in preclinical trials. Their AtomNet technology has solved more unsolved issues than any other AI drug discovery platform. Their joint ventures and collaborations with top pharmaceutical, agrochemical, and developing biotechnology businesses have a combined deal value of almost $7 billion. Atomwise has received approximately $174 million in funding from prominent venture capital companies to further the objective of developing better medications quicker.
2) PathAI [17] – Improving diagnostic pathology: PathAI creates technologies to help pathologists make a quick and accurate diagnosis for every patient, every time. PathAI is also develops technologies to assist in identifying patients who would benefit from innovative medicines, in order to make scalable customized medicine a reality.

4.3.3 Finance

1) Betterment [5–7] – Robo-adviser pioneer: Betterment is a robo-adviser that invests and administers its users' individual, IRA, ROTH IRA, and 401(k) accounts. To participate, users must complete a brief survey about their present investing situation, including questions such as whether they handle their assets themselves or use a human adviser, the amount invested, and

the type of investment. The poll also inquires about the user's tax filing status, income, debt, and yearly household income and investment plans. According to the company website, the machine learning algorithms will then utilize this information to give the user an overview of their financial situation and enable the program to propose appropriate investments.

2) Numerai [16] – An AI-powered, crowdsourced hedge fund: the Numerai Tournament requires you to create machine learning models based on abstract financial data in order to predict the stock market. Your models can be staked with the NMR cryptocurrency to gain performance-based incentives. The Numerai staked models are merged to produce the Meta Model, which handles the Numerai hedge fund's money throughout the global stock market.

4.3.4 Travel and Transportation

1) Google – Smart maps: Google Maps is Google's web mapping platform and consumer application. It provides satellite images, aerial photography, street maps, 360° interactive panoramic views of streets (Street View), real-time traffic conditions, and route planning for walking, driving, and taking public transportation. Google Maps was utilized by approximately 1 billion people per month worldwide as of 2020.
2) Hipmunk [12] – Virtual travel assistant: Hipmunk is a travel search engine that promises to simplify vacation planning. Their aim is to help consumers plan to travel more quickly and effectively. Hipmunk was created to assist users who are inundated with irrelevant search results. Hipmunk displays flight results in a graphic timeline, allowing users to choose the best flight for them at a glance. Hotel results are displayed on a map so that visitors may see where they will be staying and what sights are nearby.

4.3.5 Social Media

1) Facebook AI [10] – Image recognition breakthrough: Facebook's goal is to infer 3D shape and position from a single image, and it proposes a learning-based technique that can train from unstructured image collections while using just segmentation outputs from off-the-shelf recognition systems as supervisory signals. They begin by determining a volumetric representation in a canonical frame, as well as the camera posture for the input picture. They ensure geometric consistency with both look and silhouette, as well as that the synthesized new perspectives are unrecognizable from picture collections. The crude volumetric forecast is then transformed to a mesh-based representation, which is improved further in the expected camera frame based on the input image. These two processes enable shape-pose factorization from unannotated pictures and finer-grained reconstruction of per-instance form. They provide results from both synthetic and real-world datasets. Experiments indicate that their method captures category-level 3D

form from picture collections more correctly than alternatives and that this may be enhanced further with their instance-level specialization.

2) Snapchat [13] – Camera filter: Snapchat has swiftly introduced novel elements to the world of social media interactions, such as its Face Lenses, which are digital masks that bring a distinct type of engagement to daily images and videos through the use of augmented reality. People select lenses depending on their aims, personalities, and a scroll-first mentality. They offered one of the first examples on this new function, which has swiftly spread to various other social networking apps.

4.3.6 E-Commerce

1) Amazon [2] – AI-powered … everything: Amazon.com relies heavily on machine learning-based technologies to power most of its business. It could not expand its company, improve its customer experience and variety, or optimize its logistic speed and quality without machine learning. It created AWS to enable other businesses to benefit from the same IT infrastructure with greater agility and lower costs, and it is now working to democratize ML technology for every firm. Its development team structure and the emphasis on ML to address hard pragmatic business issues led them and AWS to build simple-to-use and powerful ML tools and services. These tools are initially tested in their own scale and mission-critical environment before being offered as AWS services for any organization to utilize, similar to other IT services.

2) Myntra (emerj) [9] – Rapid AI technology: Myntra concentrates on producing intelligent fast fashion. A modern word used by fashion retailers to imply that designs travel quickly from the catwalk to current fashion trends is "fast fashion." They also utilize Rapid to intelligently choose what to sell on the market. "You're not going to be able to sell everything. We delight in offering one-of-a-kind clothing." They use machine learning to improve payment acceptance rates for online transactions – an issue that is particularly prevalent in India, where failure rates are high. Because consumers typically complain about late refunds, they created "Sabre," an AI-based returns system that allows for speedier reimbursements for clients who have demonstrated strong buying-return behavior. According to them, by studying a customer's past return behaviors, "Sabre" can detect which consumers are truly returning packages and which are attempting fraudulent acts.

4.3.7 Marketing

1) Drift [8] – Conversational marketing: Drift is a Revenue Acceleration platform that combines Conversational Marketing and Conversational Sales to help businesses generate revenue and customer lifetime value more quickly. Drift is used by over 50,000 organizations to integrate sales and

marketing on a single platform to provide a unified customer experience in which customers are free to speak with a business at any time, on their terms.

2) Amplero [3] – Building customer relationships: Amplero is a customer lifetime value management software that helps marketers to achieve the impossible by utilizing machine learning and multi-armed bandit experimentation to improve every customer contact and increase customer lifetime value and loyalty.

4.4 INTERNET OF THINGS

IoT, commonly known as the Internet of Things, connects a device not only to the internet but also to other devices to build a network where the data in the surroundings revolves and makes human life easier. The examples discussed in the previous section for AI are intended for a particular purpose, i.e., marketing applications would be exclusive to the marketing sector only. Those applications work on the principles of the human mind, i.e., learning from the knowledge provided. Since their knowledge is limited to a particular sector or a section of society, they do not imply working for multiple sections. On the other hand, IoT applications are provided with the knowledge to connect with more than one service, i.e., the internet, to solve real-life problems and make human life easier. As per Lee and Lee [14], the IoT is a worldwide network of machines and gadgets that can communicate with one another. It does not require any substantial human interaction to function. This article covers five IoT technologies required for the effective deployment of IoT goods and services, as well as three IoT categories for business applications used to increase consumer value. Furthermore, it investigates the net present value technique and the real options approach, both of which are frequently employed in the justification of technology initiatives, and shows how the real options approach may be used for IoT investments.

Among the different applications of IoT, an suitable real-life example of the IoT is Amazon's own virtual assistant, Alexa. Initially installed in Amazon's home assistant device, Echo, Alexa has become very popular over a short amount of time. As for Alexa, other tech giants have also developed their own voice assistant that interacts with people and other services (both software and hardware) to make services more accessible and surroundings more interactive. For instance, Apple has Siri, Microsoft has Cortona, and Facebook has come up with their assistant known as "M"; Google also has its own virtual assistant.

As far as accessibility, interactivity, and ease are concerned, these services have reduced the distance between a person's outer world and private world to zero. First of all, the devices are installed in both public to private spaces. From keeping them in the living room of their house to the bedroom, which is solely considered to be a private space. Knowledge technologies that we are consuming right is more important now than ever because such technologies are also consuming us. Knowing what data access we provide to these services makes a huge difference. We have seen a lot of data breaches in recent times due to this fact. For instance, Google knows everything from our house address to our date of birth and the car we own and a lot more things that we would find meaningless, but the service and the company don't. To say that the services are a threat to our personal security is not very satisfactory since knowledge about it is important, and just like the user manual of every product, people tend to ignore that part which could make it dangerous. IoT services, as a whole, is a great field for research but the security aspect of it seems to be being left in the dark.

Abdulla et al. [1] discusses the significance of successful technology for humans. Nowadays, IoT technology plays an important part in human existence. In recent years, smart home systems and smart city security have been based on IoT technology, which supports and makes it easier to monitor appliances to enhance the availability of various devices for home automation with robust security. As a result, the Internet of Things has been exploited a major component of future wireless sensor networks that can run without human intervention.

4.5 BLOCKCHAIN

Blockchain is a Linked List that is duplicated, distributed, cryptographically linked, and kept up to date by agreement [28]. One of the primary aspects of Blockchain is the cryptographically guaranteed data integrity. It may also be used as an immutable ledger for events, transactions, or time-stamped data. The tamper-resistant log is a critical security feature that, when used correctly, makes hacking nearly impossible. It also offers a platform for creating and transacting in a cryptocurrency ledger of non-currency-related events/transactions. Blockchain technology is a digital breakthrough with the potential to dramatically affect trusted computing operations and, as a result, cyber-security problems in general.

Attractive properties of Blockchain include:

- Log of data with digital signature
- Immutable (once written – cryptographically hard to remove from the log)
- Cryptographically secure – privacy-preserving
- Provides a basis for trusted computing on top of which applications can be built

4.5.1 Cryptography

The technique for encrypting and decrypting a text segment is commonly known as a cipher. It is further divided into two types of function: encryption and decryption.

Some cryptography approaches rely on the algorithms' secrecy. These days, such approaches are primarily of historical interest. To manage encryption and decryption, all contemporary algorithms employ a key. The encryption and decryption keys, known as the public and private keys can be different (Figure 4.1).

4.5.2 Public and Private Keys

Public and private keys are essential components of cryptocurrencies developed on a Blockchain platform, which are part of a wider discipline of cryptography known as

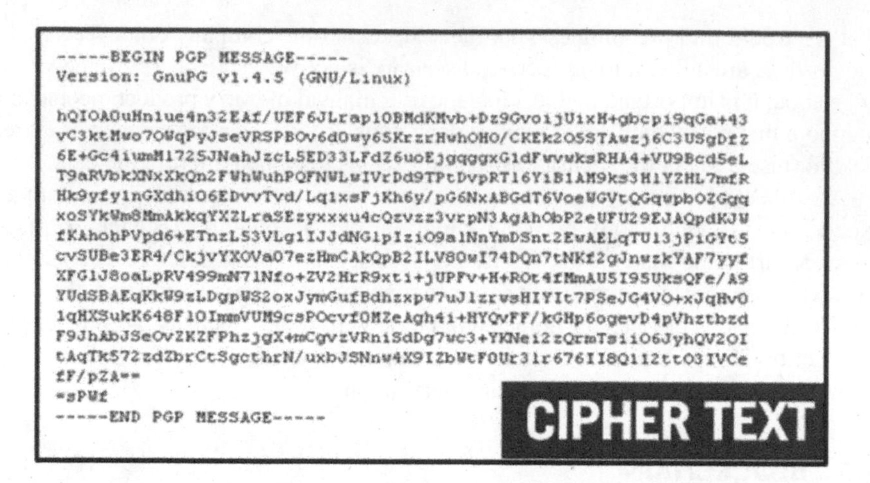

FIGURE 4.1 Cipher text.

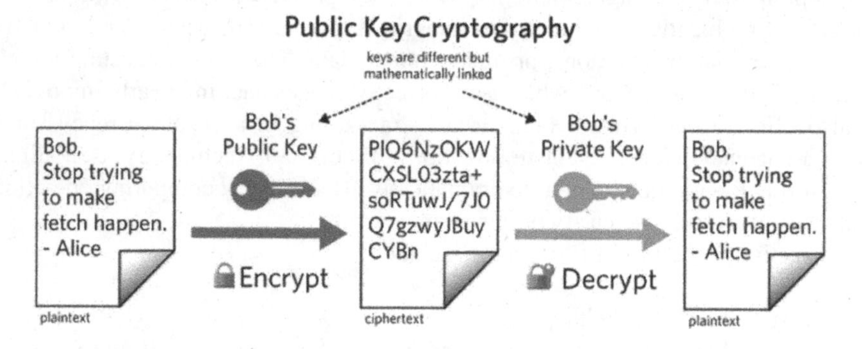

FIGURE 4.2 Public and private key.

Public Key Cryptography (PKC) or Asymmetric Encryption. A public key is used to encrypt a message that can only be decoded by the corresponding private key. Since it is a one-way mathematical function, it is excellent for confirming the validity of a transaction because it cannot be faked. PKC is based on a two-key paradigm, the public and private key, typically symbolized as a padlock, and the real key to open the same, respectively (Figure 4.2).

4.5.3 Hash Functions

A hash function takes an arbitrary length string as an input producing a fixed-size output (e.g., 256 bits). They are easy to compute and almost impossible to reverse. Security properties of a hash function are that it is collision-resistant, hides the original string making it almost impossible to get the original string from the output, and puzzle-friendly. Here we've used the SHA-256 hash function, one of the best and most commonly used (Figure 4.3).

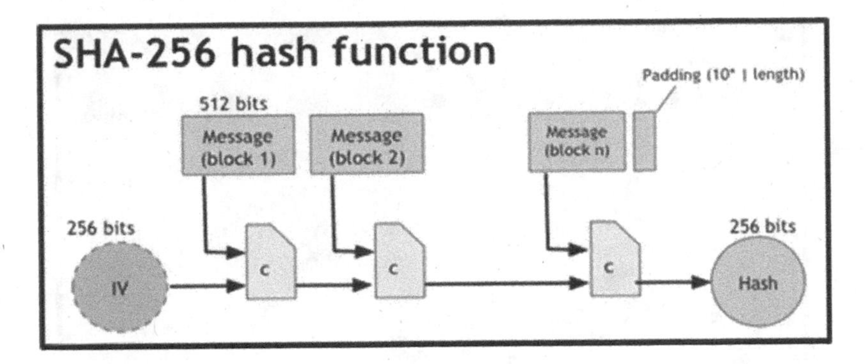

FIGURE 4.3 Sample hash function.

4.5.4 Consensus and Proof-of-Work

A distributed and trustless consensus system means that you don't have to rely on third-party services to transfer or receive money from someone. Proof-of-work is one of several forms of consensus procedure. Proof-of-work is a technique designed to discourage cyber-attacks such as distributed denial-of-service (DDoS) assaults, aiming to deplete a computer system's resources by sending repeated bogus requests. The notion of proof-of-work existed before Bitcoin, but Satoshi Nakamoto adapted it to his digital currency, changing the way standard transactions are structured. At this point, the miners must solve a mathematical challenge called a proof-of-work problem. When a miner eventually discovers the correct answer, he broadcasts it to the whole network at the same moment in exchange for a cryptocurrency award (the reward) granted by the protocol (Figure 4.4).

4.5.5 Solidity & Smart Contracts

Solidity is a high-level, object-oriented programming language used to build smart contracts. Smart contracts are programs that regulate how accounts behave within the Ethereum state. Solidity was inspired by Python, C++, and JavaScript, and it is intended for use with the Ethereum virtual machine (EVM). Solidity is statically typed, and among other things, it allows inheritance, libraries, and complicated user-defined types. Solidity allows you to build contracts for various purposes, including voting, crowdsourcing, blind auctions, and multi-signature wallets. A smart contract is a computer program or a transaction protocol designed to automatically execute, control, or document legally important events and acts according to the conditions of a contract or agreement. The goals of smart contracts are to reduce the need for trusted intermediaries, arbitration and enforcement costs, fraud losses, and deliberate and unintentional exceptions.

4.5.6 Crypto Coins and DAPPS

A cryptocurrency is a digital or virtual currency protected by encryption, making counterfeiting or double-spending almost impossible [29]. Many cryptocurrencies

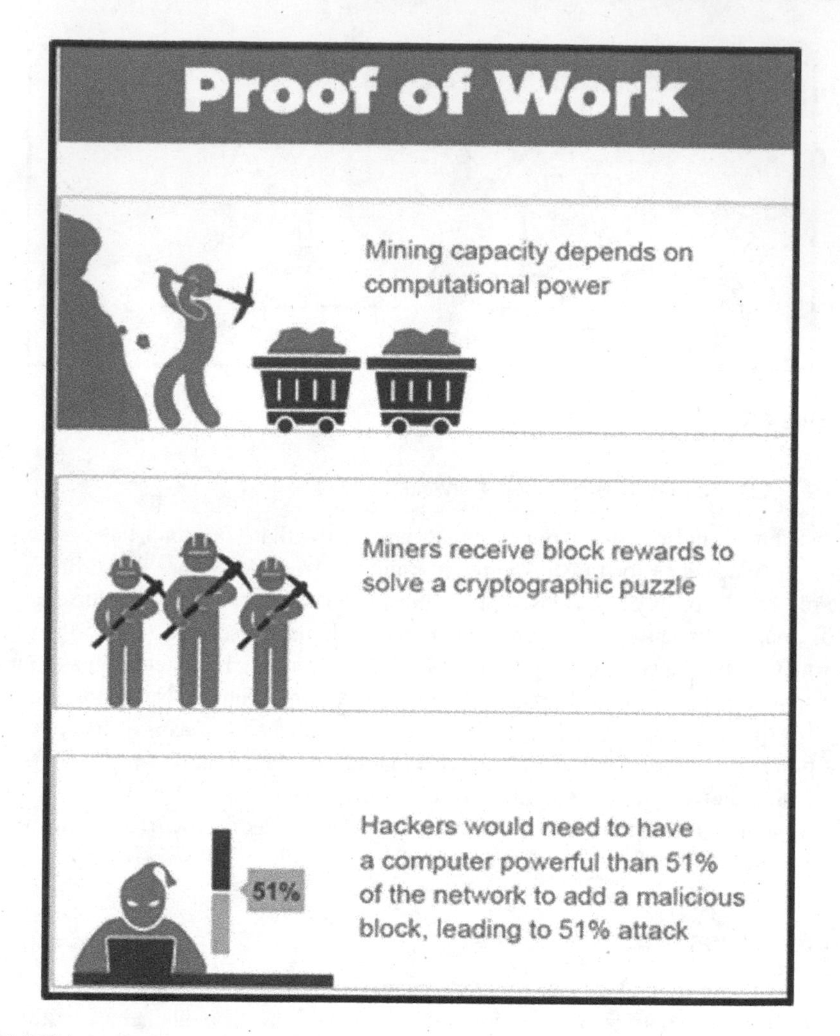

FIGURE 4.4 Proof-of-work.

are decentralized networks based on Blockchain technology, a distributed ledger enforced by a diverse network of computers. Cryptocurrencies are distinguished by the fact that they are not issued by any central authority, making them potentially resistant to government meddling or manipulation. There are many types of cryptocurrencies available, of which Bitcoin is the most valuable, followed by Etherium, and Bitcoin cash (Figure 4.5).

DAPPS stands for Decentralized Applications. DAPPS' back-end code (smart contracts) executes on a decentralized network rather than a centralized server. They use the Ethereum Blockchain for data storage and smart contracts for their app logic (Figure 4.6).

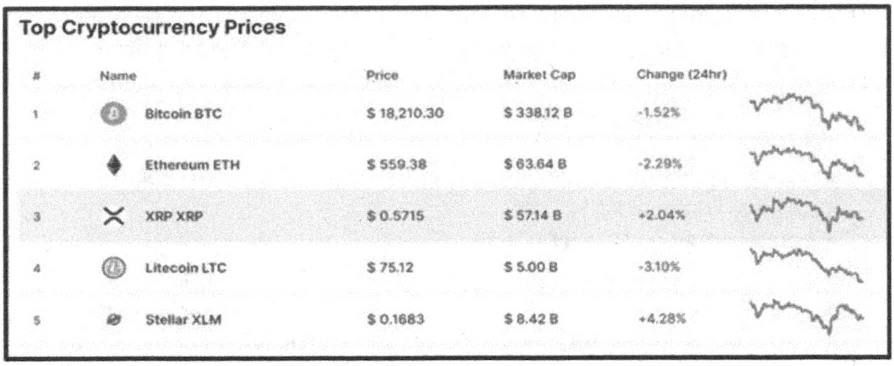

Top Cryptocurrency Prices

#	Name	Price	Market Cap	Change (24hr)
1	Bitcoin BTC	$ 18,210.30	$ 338.12 B	-1.52%
2	Ethereum ETH	$ 559.38	$ 63.64 B	-2.29%
3	XRP XRP	$ 0.5715	$ 57.14 B	+2.04%
4	Litecoin LTC	$ 75.12	$ 5.00 B	-3.10%
5	Stellar XLM	$ 0.1683	$ 8.42 B	+4.28%

FIGURE 4.5 Cryptocurrency prices.

No owners
Once deployed to Ethereum, dapp code can't be taken down. And anyone can use the dapp's features. Even if the team behind the dapp disbanded you could still use it. Once on Ethereum, it stays there.
Free from censorship
Built-in payments
Plug and play
One anonymous login
Backed by cryptography
No down time

FIGURE 4.6 Advantages of cryptography.

4.6 IoT AND BLOCKCHAIN

The idea of Tangle originated when the IOTA Foundation attempted to integrate Blockchain with the IoT [26]. Tangle is not actually a Blockchain but a technology that takes most of the advanced features of Blockchain. It is designed for communication and connection between machines and IoT devices under the surveillance of a distributed ledger. When the requirement for a self-governing IoT ecosystem arose, the most suitable option was to merge IoT and Blockchain. But it comes with a lot of challenges, as discussed below:

a) **Scalability Issues**

Blockchain works in blocks and as a transaction transpires in a block the information needs to be accounted for in real-time. The generated data needs to be scaled and kept for future reference. As the technology enhances and advances further, the amount of user-generated data will increase excessively hence making it tough for the nodes to validate all transactions, creating scalability issues.

b) **PoW Computational Requirement**

PoW is a huge power-consuming process (electricity) and requires costly computational devices that effuse high intensity of heat and temperature in a non-eco-friendly manner. In addition to this, to set up such a powerful mining structure, high expenses are required, and high manpower is also required for maintenance and keeping temperature under control using expensive cooling systems.

c) **Centralization by Powerful Miners**

People with high resource availability and expenses are able to not just set up mining structures but also centralize them using those resources. High resource availability means more power to a singular unit making it more centralized. Hence, contradicting the real purpose of Blockchain, i.e., being a decentralized self-governing technology.

d) **Cost of Transactions**

Even for small transactions, sometimes the fees to pay could surpass the actual transaction amount for the validating nodes in order to authorize the transaction so that it can enter the block. This is one of the major setbacks of the technology that stops it being taken up by more people.

e) **All Guarantees of Integrity Are Probabilistic**

Anonymity is now almost a myth, and the amount that still remains is at stake. The probability of a person being identified anonymously remains in the hands of the powerful miners who tend to centralize the whole system, as we discussed in the previous challenge. As the number of nodes under the control of a single user or organization increases, the total number of users remains approximately constant. As the control shifts from a decentralized to a centralized system, the probability of integrity reduces significantly, leading to the reduction of probability.

f) **Privacy Requires a Bit More Thought**

Since Blockchain is still under development, new kinds of suggestions and methods are introduced day in and day out to make the environment flawless. To make Blockchain more scalable and increase its integrity, soft forks and hard forks are performed on the Blockchain.

4.6.1 Features of IoT

The IoT on its own requires some basic services such as internet connectivity, sensors, and knowledge of the task it has been provided with. Once the requirements are met, the IoT works with the following features that provide them with the independence of connecting with certain machines as instructed.

1) **Low Resource Consumption**

IoT devices are highly sophisticated and efficient with simple and low resource requirements, i.e., their operation does not require any expensive services or devices; rather, they work efficiently with a simple network connection and a normal power backup.

2) **Widespread Interoperability**

Since an internet service is one of the basic requirements for an IoT device, any operation that can be executed through the internet will be applicable to the IoT device, thus making it accessible in any part of the world with a proper internet connection.

3) **Billions of Nano-Transactions**

While transacting, IoT devices do not communicate directly with the machine at the other end; instead, they send a sequence of signals that the machine interprets as nano-transactions, and the number of these transactions could be the in millions and billions. The nano transacted signals individually only take up a minute amount of memory, thus meaning they can be used in a huge number.

4) **Data Integrity**

As the name suggests, the integrity of the data in the form of nano transacted signals can only be interpreted by the machines programmed particularly for the assigned task of the device. Since the nano-transactions are channeled only between the IoT device and the corresponding machine, the chance of data breach is reduced to negligible. Thus, increasing the integrity of the data.

4.7 THE SOLUTION

The IOTA Foundation claims to have solutions for the challenges discussed above by developing the latest technology inspired by Blockchain, which originated from the same but eliminated the critical issues that the concerned technology faced. The following are the presumed solutions:

1) **Replace Blockchain with Tangle**

By replacing Blockchain with Tangle, we can eliminate the complexity and the challenges mentioned above and create a sophisticated technology that works well with IoT devices, making them easy to integrate [21–25].

2) **It Borrows a Lot of Ideas from Blockchain**

Like Blockchain, Tangle also transacts. In Blockchain, the transaction contains cryptocurrency, but Tangle transacts the information between IoT devices and machines. Here, machines have the capability to self-validate the information transmitted rather than in Blockchains where the miners validate the transactions.

4.8 TANGLE

Tangle, like Blockchain, is a technology used for various classes of transactions. A directed acyclic graph, which is similar to a distributed ledger, is employed in this case [27]. DAG is not controlled by any external body, such as a bank or financial organization. The positive aspect is that Tangle is IoT-compatible. The IoT is a network of interconnected gadgets that interact and share information. Tangle will be able to enable large transactions between multiple linked devices in a quick and seamless manner. Different features of Tangle that establish the technology are discussed below:

1) **No Block – Individual Transactions are Tangled Together**
   Like any conventional Blockchain, Tangle does not have blocks containing valid transactions; instead, it has individual transactions validated by new transactions that are tangled together.
2) **What is Tangling?**
   Tangle consists of a construct DAG connecting transactions. This is how a Tangle-based cryptocurrency operates. There is a DAG called Tangle that replaces the global Blockchain. Tangle graph's site set, which serves as the ledger for storing transactions, is made up of transactions issued by nodes. When a new transaction arrives, Tangle's edge set is formed as follows: it must first approve two previous transactions. These approvals are represented by curved edges. If there is no directed edge between transaction A and transaction B, but there is a direct path of at least two lengths from A to B, we say that A indirectly approves B.
3) **Self-Regulating**
   At the start of the Tangle, there is an address with a balance that holds all of the tokens. These coins are distributed to many additional "founder" addresses as part of the genesis transaction. Let us emphasize that the genesis transaction produces all of the tokens. In the future, there will be no tokens created, and there will be no mining in the sense that miners will receive monetary rewards. A short word on terminology: sites are transactions represented on the Tangle graph. The network comprises nodes, which are entities that issue and evaluate transactions, thus making it self-regulating.
4) **Very Scalable**
   Tangle's basic concept is as follows: users must first cooperate to approve existing transactions to issue a transaction. As a result, users that initiate a transaction help to secure the network. It is expected that the nodes verify to ensure that the authorized transactions do not contradict each other. If a node discovers that a transaction contradicts Tangle history, the node will refuse to authorize the conflicting transaction, either directly or indirectly.
5) **Still Use PoW – But a Long Overhead PoW**
   ("What is proof-of-work?") In Tangle, the whole network participates in the consensus mechanism, as opposed to Blockchains, where just a subset of the network participates in the verification process. It is crucial to remember that Tangle continues to employ proof-of-work. The load, however, is borne by the entire network. Furthermore, in Tangle, PoW is merely one component of the consensus architecture (to serve against Sybil and spam attacks). A competent scammer in Tangle's consensus process will have to expend a negligible amount of computing resources due to the nature of a DAG. A bad scammer, on the other hand, will have to use growing amounts of computing power with declining rewards.

4.8.1 What Do We Get Out of Tangle in Place of Blockchain?

1) **Consistency**

   Since the technology consists of high-level security features and self-validation features, the technology is free from any external influence, thus making the service of the Tangle technology consistent and hassle-free [30].

2) **Availability**

   Just like other technologies, such as cloud computing, as we discussed, Tangle technology can be made available across platforms and devices irrespective of systems and global boundaries.

3) **Partition Tolerance**

   Partition tolerance measures the continuity of the communication between the IoT devices even after a break in the channel that may affect the execution of the process. This means that the process of the signal transfers will become static during the partition hence picking up from where it lost the continuity. Hence, providing high partition tolerance.

4) **No Fees**

   Unlike Blockchain technology, Tangle does not carry any fees for the nano-transactions. The three main reasons behind the same are discussed below:

   a) All transactions are tangled together, and there are no blocks like in that of Blockchain, which contain only valid transactions handpicked by the validating node. The absence of the validating node in Tangle makes it free from fees.
   b) Since Tangle has nano-transactions, the new node's validation is done by making it a self-validation nano-transaction. Which again eliminates the necessity of a fee-consuming validating node. This point also contributes to the fact that Tangle is free from technology fees.
   c) As discussed in the previous sections, the count of the nano-transactions is in millions and billions, making the total charge of transactions more than that of Blockchain and also validating those transactions manually or even by validating nodes is practically impossible as it may take years to validate a single transaction.

5) **Modular**

   Modularity is a system characteristic that quantifies the degree to which tightly linked compartments within a system may be dissociated into distinct communities or clusters that interact with one another rather than with other communities. Here, Tangle exhibits the same features by providing IoT devices with separated communities of signals rather than densely connected chambers.

6) **Lightweight**

   Unlike other technologies, Tangle is not a heavily coded, high resource-demanding technology, but rather a very simplified and easy-to-use tech that supports IoT.

7) **Offline Allowed**

Like other technologies that provide offline access to the process, Tangle also provides offline transactions between the devices and the machines in the absence of internet connectivity.

8) **Quantum Proof**

("Will quantum computing break cryptocurrencies?") Quantum proof is a security feature that provides Tangle with resistance to quantum computation, i.e., the transactions cannot be revoked using any third-party quantum computations. Quantum computation cannot break cryptocurrencies and the encryption that protects them, just like in Tangle.

4.8.2 Tangle Transaction Issuance

Transaction issuance in Tangle requires some basic steps which accumulate into a simple nano-transaction. They are as follows:

1) **Bundling and Signing**

(Encryption consulting) First a transaction bundle is created. It collects all of the various transactions related to a particular value transfer intent. Your private keys are used to sign the created bundle. Data signing serves to verify the sender of the data and typically includes some sort of encryption in the process. Signing emails, sensitive data, and other material has become important since it confirms the sender's identity and assures the data has not been altered in transit. If an attack happened and the attacker changed or compromised the data receiver would be aware of this. The attacker might change the data, but because they do not have the key used by the sender to sign the data, the receiver will know not to trust the data when evaluating the key and data. Asymmetric encryption works by generating a key pair consisting of a public and private key. The private key is kept hidden from everyone except the key's author, but the public key is available to everyone. When needed, the data is encrypted with the private key and decrypted with the public key. Symmetric encryption employs a single key for encryption and decoding. Because asymmetric encryption is more secure than symmetric encryption, it is more commonly employed.

2) **Tip Selection**

When two previously refused transactions or tips are approved at random, We'll be able to join the Tangle as a result of this. A tip is a transaction that does not have any approvers. According to the Tangle protocol, all incoming transactions attach to the Tangle by approving two transactions. A critical component of the Tangle is the mechanism used to choose the tips to which a new transaction will attach itself.

3) **Validation**

Under the Tangle protocol, new transactions need to connect with at least two previous ones, validating the transactions and reducing the chance of various attacks.

4) **PoW**

(Unofficial IOTA News) A proof-of-work is a piece of data that is difficult (expensive, time-consuming) to create but easy for others to verify and fulfills certain criteria. Proof-of-work is intended to prevent spam. The IOTA ledger is secured by a broadly distributed variant of the Current Proof-of-Work protocol. IoT devices create signed transactions and Proof-of-Work in a distributed fashion.

5) **Publishing**

(iota-news) Publishing is the final step in issuing the transaction after validation and proof-of-work. It is the process of making the transaction available for the machines present in that particular Tangle ecosystem.

4.9 CONCLUSION

At first, we introduced security systems, as they have become one of the integral parts of the modern lifestyle. But the introduction of these sophisticated systems into our daily lives made our private lives more transparent and vulnerable. We can be easily tracked, surveyed, and interpreted by hacking into these systems. Like the old saying "there are two sides to every coin," if the technologies are not used under ethical guidance, they could cause great damage to humanity, as science is a very powerful tool. So increasing the security of these systems and introducing new technologies to enhance these systems are much needed.

The rise of AI in the current era is really promising if used to its full potential and can transform the world into a different dimension that cannot be explained but to be experienced in the near future. The IoT has really played an integral role in the advancement of AI. The IoT assists AI programs with a lot of information that the systems can use to build logic and achieve more accuracy. Through the development of the IoT that serves different purposes, new services are developed to make human lives easier and more accessible across the globe. A number of examples were discussed in the AI section, which when integrated with the IoT devices developed networks between the different types of machines, constructing ecosystems of their own.

Even though Blockchain is the next big invention after electricity and the internet, it is still under development. Tangle actually developed by taking inspiration from Blockchain technology. Tangle also has transactions such as DAG, and it also supports the IoT.

So by integrating Blockchain-inspired Tangle and IoT, the enhanced security of the system is multiplied, providing it with two-tier security of both Blockchain and IoT. Hence, through extensive research in the field, including subjects of security systems, Blockchain, and IoT-Tangle, we can conclude that Tangle with IoT can enhance security systems.

REFERENCES

1. Abdulla, Abdulrahman Ihsan, et al. "Internet of Things and Smart Home Security." *Technology Reports of Kansai University*, vol. 62, no. 05, 2020, p. 12.

2. Amazon.com. "What is Artificial Intelligence?" *Amazon*, https://aws.amazon.com/machine-learning/what-is-ai/
3. Amplero. "Amplero." *Crunchbase Company*, https://www.crunchbase.com/organization/amplero
4. Atomwise. "Atomwise." *Atomwise*, https://www.atomwise.com/company/
5. Betterment. "Robo-advisors and Artificial Intelligence – Comparing 5 Current Apps." *emerj*, https://emerj.com/ai-application-comparisons/robo-advisors-artificial-intelligence-comparing-5-current-apps/
6. Bhandary, Mohan, et al. "A Blockchain Solution based on Directed Acyclic Graph for IoT Data Security using IoTA Tangle." Fifth International Conference on Communication and Electronics Systems, 2020, https://ieeexplore.ieee.org/abstract/document/9137858
7. Carmo, Jose. "Publishing Data in IOTA Tangle with OutSystems." *IOTA News Unofficial*, https://iota-news.com/publishing-data-in-iota-tangle-with-outsystems/
8. Drift. "Drift." *Crunchbase Company*, https://www.crunchbase.com/organization/drift
9. emerj. "Artificial Intelligence at India's Top eCommerce Firms." *emerj*, https://emerj.com/ai-sector-overviews/artificial-intelligence-at-indias-top-ecommerce-firms-use-caes-from-flipkart-myntra-and-amazon-india/
10. FACEBOOK AI. "Shelf-Supervised Mesh Prediction in the Wild." *FACEBOOK AI*, 2021, https://ai.facebook.com/research/publications/shelf-supervised-mesh-prediction-in-the-wild/
11. Hanson Robotics. "Hanson Robotics." *Hanson Robotics*, https://www.hansonrobotics.com/about/
12. Hipmunk. "Hipmunk." *Crunchbase*, https://www.crunchbase.com/organization/hipmunk
13. "How Users Choose a Face Lens on Snapchat." *researchgate*, https://www.researchgate.net/publication/328679212_How_Users_Choose_a_Face_Lens_on_Snapchat
14. Lee, In, and Lee Kyoochun. "The Internet of Things (IoT): Applications, Investments, and Challenges for Enterprises." *Business Horizons*, vol. 58, no. 4, 2015, pp. 431–440.
15. iRobot Corporation. "iRobot." *Wikipedia*, https://en.wikipedia.org/wiki/IRobot
16. Numerai. "Numerai Tournament." *Numerai Tournament*, https://docs.numer.ai/tournament/learn
17. PathAI. "PathAI." *PathAI*, https://www.pathai.com/what-we-do/
18. Popov, Serguei. "The Tangle." 2018, p. 28, http://www.descryptions.com/Iota.pdf
19. "Publishing Data in IOTA Tangle with OutSystems." *iota-news*, https://iota-news.com/publishing-data-in-iota-tangle-with-outsystems/
20. Toch, Eran, et al. "The Privacy Implications of Cyber Security Systems: A Technological Survey." 2018.
21. upGrad. "Tangle vs Blockchain." *upGrad Blog*, https://www.upgrad.com/blog/tangle-vs-blockchain/
22. Valouch, Jan. *Integrated Alarm Systems*. Springer, Berlin, Heidelberg, 2012, https://link.springer.com/chapter/10.1007/978-3-642-35267-6_49#citeas
23. "Will Quantum Computing break Cryptocurrencies?" *FORBES*, https://www.forbes.com/sites/rogerhuang/2020/12/21/heres-why-quantum-computing-will-not-break-cryptocurrencies/?sh=40b1bb05167b
24. "IOTA Proof-of-Work: Remote vs Local Explained." *IOTA News Unofficial* https://iota-news.com/iota-proof-of-work-remote-vs-local-explained/
25. "What is the Difference between Encryption and Signing?" *Encryption Consulting* https://www.encryptionconsulting.com/education-center/encryption-and-signing/
26. Silvano, Wellington Fernandes, and Roderval Marcelino. "Iota Tangle: A Cryptocurrency to Communicate Internet-of-Things Data." *Future Generation Computer Systems*, vol. 112, 2020, pp. 307–319.

27. Guo, Fengyang, et al. "Characterizing IOTA Tangle with Empirical Data." *GLOBECOM 2020–2020 IEEE Global Communications Conference*. IEEE, 2020.
28. Yaga, Dylan, et al. "Blockchain Technology Overview." *arXiv* preprint arXiv:1906.11078 (2019).
29. Pilkington, Marc. *Blockchain Technology: Principles and Applications*. Research Handbook on Digital Transformations. Edward Elgar Publishing, 2016.
30. B. Shabandri, and P. Maheshwari, "Enhancing IoT Security and Privacy Using Distributed Ledgers with IOTA and the Tangle," *2019 6th International Conference on Signal Processing and Integrated Networks (SPIN)*, 2019, pp. 1069–1075, doi: 10.1109/SPIN.2019.8711591

5 Recent Trends in 5G and Machine Learning, Challenges, and Opportunities

S. Kannadhasan, R. Nagarajan, and M. Shanmuganantham

CONTENTS

5.1 INTRODUCTION

Antenna arrays and quadrille helical antennas are two kinds of antennas that have been tested. At particular high frequencies, beamforming methods, which produce 3D beams that direct broadcast power to particular users, enhance antenna gain, and therefore improve receiving power, which may compensate for propagation and penetration losses. Because it takes extensive domain expertise and is time and labor expensive, the existing practice of developing and upgrading antennas by hand has a limited capacity for manufacturing new and upgraded antenna designs. This approach employs evolutionary algorithms (EAs), which are a type of stochastic search technique inspired by natural biological evolution, that work on a population of possible solutions and leverage the test theory's endurance to build better and better approximations to a solution. Due to the complexity of electromagnetic interactions, evolutionary algorithms have been used to build antennas in-situ, that is, taking into consideration the impacts of surrounding structures, which is very difficult for antenna makers to accomplish by hand [1–5].

Furthermore, beam creation may aid users in limiting disturbance by using the mmWave channel's unique propagation properties, such as spatial sparsity. These high objectives need a whole new antenna design and optimization method. Antenna

DOI: 10.1201/9781003229704-5

arrays should no longer be planned and configured without taking network topology into consideration: in addition to classic antenna architectural aims like decreasing side-lobe levels and increasing directivity, additional global, network-oriented criteria must be considered [6–10]. The difficulty is significantly more when the optimization objective changes from a solely electromagnetic to a networked world. For most enterprises and universities, antenna development and network implementation experiments are too costly. The use of electromagnetics and network simulations is common [11–15]. The need for proper antenna modeling was shown in network simulators, demonstrating that design and optimization must be done simultaneously. In certain circumstances, heuristic simulation-based optimization is not possible; as a consequence, complicated simulators are time and computationally costly. Because optimization techniques need a huge number of iterations, simulations that take hours (or even days) to perform are not viable. We present and test a machine learning (ML) framework in this article that can mimic a given simulator and help us achieve any network optimization objective in a reasonable period of time [16–20].

Cell phones and the internet are used to connect people. Enablers for the twenty-first information society (METIS), millimeter-wave evolution for backhaul and access (MiWEBA), and other enablers for the twenty-first information society channel models are among the various 5G wireless communication channel formats [21–25]. Both deterministic and stochastic channel models exist, such as the ITU-R IMT-2020 channel model, IEEE 802.11 channel models, the millimeter-wave based mobile radio access network for fifth-generation integrated communications (mmMAGIC) channel model, quasi deterministic radio channel generator user manual and documentation (QuaDRiGa) channel model, and a general three-dimensional (3D) nonstationary 5G channel model. MIMO and mmWave, two of the most crucial 5G cellular networking technologies, have received a lot of attention. The data from massive MIMO and mmWave indoor channels was used to derive major MIMO and mmWave channel characteristics. In 5G cellular networks, two types of AI designs have been employed so far, as shown in Figure 5.1. To preprocess measurement findings, one technique is to employ mathematical learning

FIGURE 5.1 5G technologies in machine learning.

procedures such as clustering algorithms. When determining density, the proposed Kernel–Power–Density technique employs the kernel density and only considers the neighboring points [26–30].

5.2 MACHINE LEARNING

A unique clustering design that takes elevation angles into consideration was created using the Kernel–Power–Density method. As a solution to the tracking problem, the Kuhn–Munkres algorithm was presented. To keep track of the clusters and predict their locations, the Kalman filter was utilized. In addition, various additional techniques, such as the K-Power Means method and the hierarchical tree, were employed to categorize clusters in measurement data preprocessing. In traditional cluster-based stochastic channel models, the above-mentioned clustering techniques play a crucial role. Another option is to anticipate channel characteristics using machine learning algorithms by analyzing the mapping link between physical environment data and channel characteristics. Two kinds of artificial neural networks (ANNs), namely multilayer perceptron (MLP) and radial base function (RBF), were used to represent the link between frequency, height, and path loss (PL). PL was represented in MLP as a mapping connection between delay and climate. RBF and MLP were employed to simulate Doppler frequency transmission. A feed-forward network (FFN) and a deep architecture were used to describe the mapping connection between channel attributes and geographical position.

An FFN and an RBF network with a three-layer structure based on ML – "wave, cluster-nuclei, and channel"– were utilized to mimic in-vehicle wireless channels at 60GHz. The mapping link between a single channel feature and knowledge of the physical channel backdrop is the sole way to see detailed channel features in most current investigations. Simultaneously, given the importance of channel characteristics in any sub-channel in a given context for channel estimation and communication dependability, it is an oversight that they have yet to be estimated. Because CNN is effective at compressing and evaluating redundant channel findings, it hasn't been utilized to anticipate channel characteristics. We use CNN to derive the mapping connection between transmitter (Tx) and receiver (Rx) antenna position information and practically all amplitude, latency, and angle properties in this study.

Traditional antenna designs rely on calculations and practice to modify the necessary parameters on a regular basis, which is time-consuming and stressful. In antenna design, genetic algorithms are often employed as a machine-learning technique to search for large-scale, nonnutritive solution space and choose the optimal parameter value. A genetic algorithm was used to optimize the shape and duration of a wire antenna. An enhanced hierarchical Bayesian optimization approach was used to optimize the antenna array feed network. The single-objective evolutionary algorithms utilized in the above-mentioned research, however, may not be ideal for real-world scenarios since they overemphasize the significance of one measure. To overcome this problem, we employ a multiobjective evolutionary algorithm using gain, side-lobe, return loss, and voltage standing wave ratio (VSWR) as objective functions to optimize the antenna.

Machine learning is a relatively new method with applications in a variety of domains such as engineering, physics, psychology, and economics, to mention a few. Computational statistics is a branch of artificial intelligence that focuses on developing a mathematical model to describe input and output outcomes. A data-driven paradigm is a mathematical construct that may be used instead of scientific evidence to arrive at conclusions. The data-driven model can tackle regression difficulties because it can interpolate output depending on unknown facts. Machine learning has been extensively used for classification issues such as image identification, voice recognition, and target categorization, since classification issues have the same mathematical foundation as regression. Machine learning has recently been employed in the field of electromagnetics (EM).

Many antenna parameter optimization and radar objective categorization issues might be addressed thanks to advances in machine learning techniques, notably deep-learning technology, which permits the modeling of high-level abstractions in data. Machine learning techniques can address antenna optimization, which has a significant computational complexity. This reduces the amount of time required. Machine learning, such as that utilized in artificial neural networks, will be employed instead of a computationally costly EM simulator. Moreover, radar-based target categorization has been a burning issue in security and surveillance machine learning for decades. Machine learning applications in the EM area, especially antenna construction and radar target categorization, are discussed in this article.

Basic machine-learning and deep-learning ideas are now presented, as well as their applications in antenna geometry parameter optimization and target categorization using radar imaging. Machine learning may be used to describe a system using mathematical equations when an analytical model of the system is absent and input and output data can be analyzed or simulated. A number of parameters in the mathematical model may be optimized to approach the unit transformation function. Preparation is the process of deciding on a parameter and includes looking for parameters that will allow the model to balance the data as closely as possible. The neural network is a well-known mathematical model that has long been a part of the data-driven paradigm of machine-learning algorithms. The neural network is organized into levels, with each layer containing several experiences. Through weighting considerations, one layer's expectations are connected to those of another layer. The activation function is a non-linear function that is added to the perceptron to help with the non-linear description of the system. Deep learning differs from traditional machine learning in that it employs many layers to aid abstraction and generalization.

Traditional neural network utilization demands a feature extraction method prior to usage in order to lessen the network's computation load. If the characteristics are to be retrieved inside the neural networks, a high network size and robust training capabilities are necessary. The odds of efficient training are modest given realistic computer power and memory space. Large networks may now be efficiently trained thanks to recent increases in CPU power and memory capacity. A single deep-learning network can extract features, classify them, and generalize them, resulting in significant gains. Deep neural networks (DNNs), deep convolutional neural networks (DCNNs), and deep recursive neural networks (DRNNs) are only a few examples of sophisticated deep-learning frameworks. DCNN has gained a lot of traction in image identification, whereas DRNN has shown its worth in time-domain signal processing.

5.3 NEURAL NETWORK

Data may be sent at speeds of hundreds of megabits per second. On the other hand, planar UWB has the disadvantage of being determined by a time-consuming trial-and-error approach that requires a very complicated full-wave electromagnetic simulation. Quick model equations that offer a reasonable approximation to the final design are desirable while designing antennas. This topic is discussed in this study, and two novel soft measuring methodologies for finding the principal parameters are proposed. Two comparable ANN-based strategies are employed to construct a rectangular patch antenna in this article. Back-propagation is utilized in the first, while radial base functions are utilized in the second. Back-propagation is a useful method in a variety of situations. It may be used to solve pattern recognition and fitting issues. Due to the usage of a tapped delay line, it may also be utilized to solve calculation difficulties. The notion of structural approximation underpins the idea of a radial base function network. The network's output is a linear blend of the inputs' radial basis functions and neuron parameters. Approximation, timing, grouping, and interface control are just a few of the uses for radial base function networks. Due of the neural network's various attractive properties, these two techniques were employed in this article to simulate the connection between the rectangular planar antenna's parameters and the determined resonant frequency output. After that, the outcomes of the two tactics are compared, and the contrasts between them are fully investigated.

The ANN technique is utilized to optimize the UWB antenna in this work. It is preferable to use a feed-forward network with a RBF. ANNs, on the other hand, are openly modeled dispensing techniques (algorithms) based on the synapse development of the cerebral mammalian cortex but in a smaller size. A large ANN model may contain hundreds or thousands of processing units, but a human brain has billions of neurons, resulting in a significant increase in the breadth of their overall interaction and development behavior. When all the neurons in a network are working together, the whole behaviour of the network has energy. The neural network begins to grow: the neurons assess their output by taking into consideration the inputs; a weighted quantity is generated and assessed to a level to see how competent they are. This concurrent system is very complicated, and trying to isolate individual neurons is not possible. Artificial neural networks are the greatest option for constructing a statistical microwave circuit design (ANNs). In terms of numerical efficiency, neuro models beat EM models. After learning data from a "fine" model was utilized to train them using either EM simulation or measurement, neuro models were able to accomplish accurate and stable optimization as well as design in the testing stage. ANN was used for microwave design, simulation, and optimization.

In ANN investigation, the synthesized resonant frequency is compared to the desired resonant frequency. Finally, this study uses ANN to suggest a generalized architectural technique for micro strip antennas, which is then evaluated using the patch paradigm. Multiple-input multiple-output (MIMO) is a major technology enabler for future mobile networks to compete with the ever-increasing demand for data-hungry applications (MIMO). Through a MIMO base station with a broad range of antennas, individual information beams may be provided to numerous users at the same time with little inter-user interference. When appropriately beamforming architecture is implemented, MIMO gives not only superior spectrum accuracy and power per area, but also higher energy quality. Antenna selection (AS) is used to maximize MIMO performance in terms of both economic and technical variables as the number of antennas in MIMO grows, because radio frequency (RF) chains are often more costly than antenna components. A good AS approach, for example, may provide optimal spatial variety while drastically lowering the energy consumption of the RF chains, enhancing the device's energy efficiency. In general, AS is an NP-hard issue for which the optimum solution can be found only by scanning all possible antenna combinations exhaustively.

Its practical usefulness will be limited due to its great complexity, particularly in 5G networks that need low latency and real-time decision-making. Machine learning in communication networks has received lot of attention lately. A key benefit of machine learning-assisted communications is the capacity to define basic connections between system parameters and goals, allowing the computational load of real-time computing to be moved to the offline training cycle. A multi-class classification solution to the AS problem in single-user MIMO systems is based on two classification methods: multi-class k-nearest neighbors and support vector machine (SVM). A neural network-based solution for decreasing the numerical complexity of AS for multicasting. The neural network (NN) is used to anticipate which antenna subset will provide the best signal-to-noise ratio for the user. To increase the wiretap channel's safety, the researchers developed a learning-based broadcast antenna range. We looked at two learning-based SVM and naive-Bayes systems. There is just one antenna method accessible if it is possible to increase secrecy performance by decreasing input overhead.

Because antenna design needs the use of a computationally costly EM simulator, simulation timeframes will vary greatly depending on antenna size, operating frequency, and processing capability. Rapid optimization approaches are very useful since the antenna must be optimized to maximize performance while reducing size. Machine learning algorithms may be used instead of EM simulators to replace general optimization techniques like genetic algorithms, particle swarm optimization, and virtual annealing. The relation between patch antenna size and resonance frequency, for example, is trained using neural networks, enabling the size to be assessed quickly. Another option is to use a machine-learning methodology to estimate the parameter for the antenna parameter model. Once the model parameters are defined, the antenna properties, which may be defined by a mathematical model, may be calculated. Neural networks have recently been used to construct array antennas.

The identification and recognition of the measured target is one of the goals of radar signal processing. Goal detection is achievable because of the qualities mentioned above, such as RCS, range profile, and radar imaging. Object identification issues have traditionally been solved using classifiers based on machine-learning methods such as neural networks and support vector machines. DCNN has lately gained a lot of attention for image processing issues, and its success has led to its use in radar imaging. Fully linked layers have a class border, while convolution filters separate information in a single neural network. As a result, DCNN has been used to analyze micro-Doppler signals. Radar is used to track human gestures, hand movements, aquatic activities, drones, and automobiles, and the resulting spectrogram is categorized using DCNN. Target tracking using synthetic aperture radar (SAR) pictures has also made substantial advances. Not only did this enhance SAR image identification, but it also allowed DCNN to take charge of the whole process instead of depending on the typical detection, bias, and classification approach.

The lack of training results is one of the major drawbacks of employing deep learning for radar imaging. The number of photos required to train big networks is insufficient since the cost of collecting radar pictures is higher than that of acquiring optical camera photos. Transfer learning has been proposed as a solution to this problem, and it has shown to be quite effective. Only the last level, the fully connected layer, has to be retrained using pre-trained convolution neural networks like AlexNet and VGG16. In its most basic form, a monopole antenna consists of a radiating patch on one side of a dielectric substrate and a ground on the other. Patches made of copper or gold may be cut into any form you like. For simplicity of execution and analysis, general kinds such as square, rectangular, and circular frameworks are often utilized. Due to their basic geometry and minimal deterioration of their radiation pattern within their impedance bandwidth, a number of regular forms favor rectangular and square monopoles.

5.4 APPLICATIONS OF MACHINE LEARNING AND 5G

The Internet of Things (IoT), which connects a broad range of devices, is a critical component of 5G mobile communications. Traditional backscatter communication techniques, on the other hand, rely on the reader being able to affect radio frequencies, thus IoT deployment is restricted owing to low power supplies. Ambient backscatter communication, unlike typical backscatter communication (for example, passive sensors and RF identification (RFID) tags), does not need specialized equipment to supply power, instead using RF waves in the environment as both energy and signal resources for reflection. As a consequence, ambient backscatter allows for long-term and self-sustaining connections while lowering device maintenance and deployment expenses. We picked 4G, 5G, and Wireless-Fidelity (Wi-Fi) signals with frequencies in the range of 2–4GHz as the ambient resources since the ambient setup does not need extra spectrum services. There are still a slew of obstacles to overcome. Wireless signals may be used by unauthorized eavesdroppers to acquire information material, and transmissions of the same frequency are overlaid at the receiver to cause interference, making signal identification difficult. As processing power has expanded

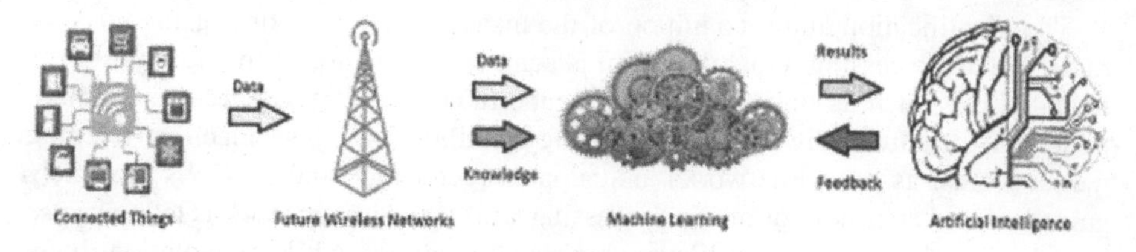

FIGURE 5.2 Applications of 5G in machine learning.

dramatically, traditional encryption techniques that encrypt data using computationally demanding codec algorithms have progressively failed. The primary idea behind physical layer security is to use the noise's inherent unpredictability to your advantage. The frequency spectrum of ultra-wideband (UWB) is made up of ultra-short bursts in the frequency domain. Working with UWB has the primary goal of obtaining high data rates while being consistent with wireless networking standards.

We investigated the application of deep learning in applied EM. Advances in machine-learning algorithms might aid other fields, such as engineering. Since there has been no visible progress in the radar image classification, the use of the radar image classification in the antenna design field is intended to enable non-experienced engineers by aiding in the design of conventional antennas or discovering innovative antenna designs. In order to get high numerical performance, the optimization of architectural parameters is currently the most significant problem, as shown in Figure 5.2. The literature has suggested genetic algorithms (GA), particle swarm optimization (PSO), Biogeography Based Optimization (BBO), neural networks, and other approaches for design optimization. The optimization methodology is a method for determining the minimum and maximum of a cost function-defined operator. The optimizers alter the vector equations before the lowest has a firm hold on anything. The error function (EF) or search methods (SM) formulas identify the optimizers. The majority of the above-mentioned optimization techniques may be applied in simulation applications. The electromagnetic simulation software HFSS and CST employ one or more of the mathematical optimizers listed below to obtain a broad variety of capabilities: Non-linear programming approaches include ANN (MATLAB compatible), Quasi Newton, Sequential Non-linear Programming (SNLP), Sequential Mixed Integer Non-linear Programming (SMINLP), and GA. ANN allows you to customize antenna characteristics including return loss, bandwidth, scale, and gain. To increase energy economy while preserving users' quality of service (QoS) needs, a collaborative architecture for antenna selection and power delivery for multi-user multi-antenna downlinks was developed.

The optimal antenna subset is chosen in a future round. Theoretically, JASPA convergence is assured. Second, to address JASPA's combinatorial complexity, which might restrict its usage in real-time applications, a machine learning-based antenna selection and power allocation (L-ASPA) approach is proposed. L-ASPA employs the NN advance to allow offline training of the underlying link between system parameters and antennas. A single-user MIMO technique is used for the multi-user technique. Because we examine universal unicast rather than complete broadcasting without inter-user interference, our suggested L-ASPA is fundamentally different. The 5G networks are expected to have ultra-high data transmission (about 10Gbps), which is approximately 1,000 times faster than current LTE networks, connect enormous numbers of people (approximately 10–100 times the size of present mobile phones), have ultra-low latency (about 1ms), which is approximately five times smaller than current LTE networks, and be incredibly energy efficient, with ten times longer battery life. They should be able to meet the diversified needs of multiple networks, such as enhanced mobile broadband (eMBB), massive machine type connection (mMTC), and ultra-reliable and low latency communication, in order to create a fully connected smart world (URLLC). Because these networks need distinct kinds of network technology, they cannot be effectively dispersed across a single homogenous network (for example, eMBB services need incredibly high bandwidth and mMTC services need ultra-dense connectivity). Because of the use of network function virtualization (NFV) and software defined networking (SDN) technologies, 5G networks may also create network slices with adequate capacity on demand. The process of dynamically configuring and managing a network slice is known as network softwarization. It is necessary to process a significant amount of complicated data.

Automating network slicing activities are simple thanks to machine-learning approaches. Sensing (e.g., anomaly detection), mining (e.g., service categorization), forecasting (e.g., predicting customer or traffic patterns), and inference are all examples of machine learning (e.g., configuration of system parameters for adaptation). It might, for example, evaluate a large amount of data in a short amount of time, learn to adapt the system to changing scenarios, generate relatively accurate predictions of future occurrences, and provide constructive solutions. Network control and management services that have been suggested for automation include resource management, on-demand and scalable network setup, infrastructure development and orchestration, fault detection, security, mobility management, user experience improvement, and dynamic policy modification. AI and deep-learning-based network control and management applications are now possible because to big data, cloud computing, cyber physical systems (CPS), and the IoT.

5.5 CHALLENGES IN MACHINE LEARNING

New 5G networks have expanded vulnerabilities and privacy issues beyond the mobile device to the service provider network by increasing capacity, spectrum utilization, and high data rates. As a consequence, the network may be able to deal with these challenges in real time, and machine learning and artificial intelligence techniques may be able to help model these highly complicated algorithms that can recognize network faults and provide a viable solution in real time. Similarly, AI and machine learning defend the network by providing adaptable protection mechanisms that can deal with a wide range of network events, threats, and attacks. In the short to medium future, AI and machine learning might be utilized to detect threats and respond with strong and scalable defense methods. A fully integrated defense system for quick response to threats and assaults is envisioned in the long term. When compared to its predecessors, 5G networks are projected to support even more heterogeneity (in terms of linked devices and networks). For example, 5G networks enable self-driving vehicles, smart homes, smart buildings, and smart cities. Similarly, the IoT may employ more reliable and scalable techniques to deal with sensitive security challenges on both the network and system levels in a 5G network topology.

The defense of such networks would be significantly more challenging due to both foreign and local interference. AI and ML may be able to help by identifying susceptible protection links in the middle, such as identification, authentication, and assurance. The security and safety of 5G-IoT would cover all levels, including identity, privacy, and E2E compliance. When hiding the key identifier, the primary authentication framework from end-device to core network and forward to service provider, for example, is still a challenging challenge to address. We believe that artificial intelligence and machine learning will aid in key authentication as well as effectively decreasing masquerade attacks.

It becomes a time-consuming procedure to ensure the security and privacy of data from these numerous networks, each with its own set of security requirements. Powerful AI and ML combined with a study of SBA and protection standards for different end-systems will find and correct these errors in real time by identifying and grouping rare threats. As a consequence, the labor shortage in the computer security industry would be greatly reduced. AI and ML can contribute to the establishment of security frameworks for the whole 5G-IoT network by establishing confidence models, application protection, and data assurance.

Fifth-generation (5G) wireless networking networks are approaching completion. Ultra-fast and stable internet connections are possible with 5G. By 2023, the worldwide cell data traffic will have surged to 100 exabytes per month, almost double the current amount of 31.6 billion mobile devices. In possible 5G networks, the infrastructure complexity in terms of network design and cellular connectivity will greatly improve. On the other hand, the amount of open resource available to each user/average device would be relatively limited.

As a consequence, network traffic management and optimization will face significant problems due to the fast expansion in data volume and consumer devices. Current research on 5G network traffic management is also pushing the limits of classic information theory-based solutions. It will be very difficult to solve the traffic management difficulty for 5G networks and provide worldwide optimal coverage for the whole network. This necessitates the development of innovative ideas. Using artificial Intelligence (AI) to monitor and manage traffic in 5G networks based on network data is a potential solution to resolve the issues stated above. AI technologies can not only reduce human interventions in network traffic management by learning new lessons from networks and anticipating network traffic dynamics and user behavior, but they can also increase network efficiency, stability, and flexibility, enabling wiser choices to be made autonomously.

In the 5G era, the utilization of cellular networks will enhance dramatically, resulting in a large increase in traffic outcomes. As a consequence of the exponential rise of content-centric apps, data service by information collecting systems has become a substantial consumer of traffic on the mobile internet (e.g., social networking applications). Data is often processed in a consolidated DC on today's networks.

TABLE 5.1
Number of Papers Published on Machine Learning

Year	Number of Papers Published					
	IEEE	IET	ELSEVIER	SPRINGER	WILEY	ACM
2010	20	20	10	30	20	10
2011	30	30	20	35	30	20
2012	50	35	30	40	50	30
2013	60	40	40	45	60	40
2014	70	48	45	50	70	45
2015	80	55	60	55	80	60
2016	100	60	70	59	20	70
2017	120	78	75	70	30	75
2018	150	100	60	99	35	60
2019	160	150	80	58	40	80
2020	180	120	86	100	48	86

For all access, data must be transferred from the DC to the UE through the core network and the radio access network (RAN). To minimize latency and relieve load on core network traffic in the 5G network, data, including recommended data, will be stored first in a local DC's mobile cloud engine (eMBB scenario). The traffic characteristics of content-centric apps would move from persistent core network access to random core network access in this scenario, which is outside the purview of current core traffic control research.

The importance in the field of machine learning research for various journal publications in the years 2010–2020 (Table 5.1), and its Comparison Chart Publication in 2010–2020, is shown in Figure 5.3

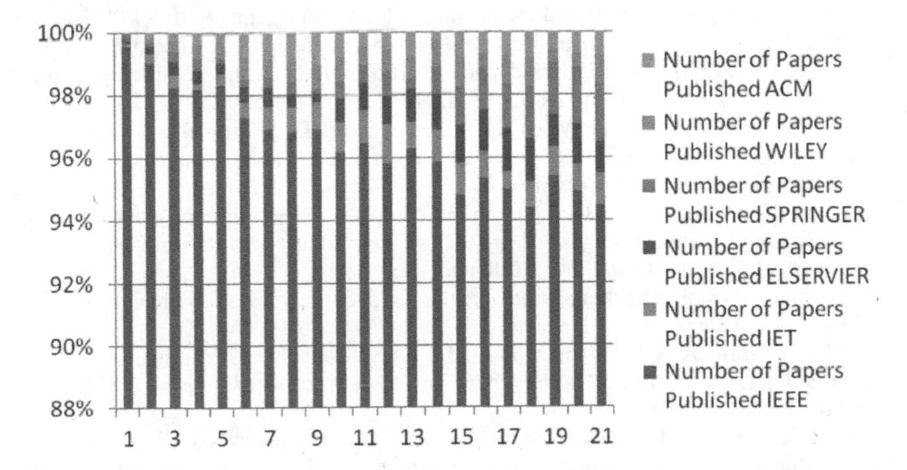

FIGURE 5.3 Comparison Chart Publication in 2010–2020.

5.6 CONCLUSION

The relevance of network feature automation in network slice preparation, growth, deployment, operation, control, and management is shown. We'll next look at a few deep-learning strategies for automating network operations. We also look at how artificial Intelligence and machine learning are used in different standards development organizations (SDOs) and commercial networks. The major contribution of this study is to highlight the relevance of network slicing functions and to look at how machine learning may be utilized to provide autonomous 5G slicing operations. When it comes to integrating common technologies for automating 5G network slicing operations using machine-learning approaches, this topic would be an excellent place to start.

REFERENCES

1. I. Giannakis et al., "A machine learning-based fast-forward solver for ground penetrating radar with application to full-waveform inversion," *IEEE Trans. Geosci. Remote Sens.*, https://doi.org/10.1109/TGRS.2019.2891206.
2. Z. Wei and X. Chen, "Deep-learning schemes for full-wave nonlinear inverse scattering problems," *IEEE Trans. Geosci. Remote Sens.*, https://doi.org/10.1109/TGRS.2018.2869221.
3. Z. Liu et al., "Direction-of-arrival estimation based on deep neural networks with robustness to array imperfections," *IEEE Trans. Antennas Propag.*, vol. 66, no. 12, pp. 7315–7327, Dec. 2018.
4. S. Chen et al., "Target classification using the deep convolutional networks for SAR images," *IEEE Trans. Geosci. Remote Sens.*, vol. 54, no. 8, pp. 4806–4817, Aug. 2016.
5. Y. Yang et al., "DECCO: Deep-learning enabled coverage and capacity optimization for massive MIMO systems," *IEEE Acc.*, vol. 6, pp. 23361–23371, 2018.
6. X. Wang et al., "BiLoc: Bi-modal deep learning for indoor localization with commodity 5GHzWiFi," *IEEE Acc.*, vol. 5, pp. 4209–4220, 2017.
7. Y. Kim and Y. Li, "Human activity classification with transmission and reflection coefficients of on-body antennas through deep convolutional neural networks," *IEEE Trans. Antennas Propag.*, vol. 65, no. 5, pp. 2764–2768, May 2017.
8. D. Shi et al., "A novel method for intelligent EMC management using a 'knowledge base,'" *IEEE Trans. Electromag. Compat.*, vol. 60, no. 6, pp. 1621–1626, Dec. 2018.
9. T. OShea and J. Hoydis, "An introduction to deep learning for the physical layer," *IEEE Trans. Cog. Commun. Netw.*, vol. 3, no. 4, pp. 563–575, Dec. 2017.
10. H. Sun, X. Chen, Q. Shi, M. Hong, X. Fu, and N. D. Sidiropoulos, "Learning to optimize: Training deep neural networks for wireless resource management," in *IEEE Int. Workshop Signal Process. Adv. Wireless Commun.*, Jul. 2017, pp. 247–252.
11. L. Lei, T. X. Vu, L. You, S. Fowler, and D. Yuan, "Efficient minimumenergy scheduling with machine-learning based predictions for multiuser MISO systems," in *Proc. IEEE Int. Conf. Commun.*, May 2018, pp. 1–6.
12. J. Joung, "Machine learning-based antenna selection in wireless communications," in *IEEE Commun. Lett.*, vol. 20, no. 11, pp. 2241–2244, Nov. 2016.
13. M. S. Ibrahim, A. S. Zamzam, X. Fu, and N. D. Sidiropoulos, "Learningbased antenna selection for multicasting," in *Proc. IEEE Int. Workshop Signal Process. Adv. Wireless Commun.*, Jun. 2018, pp. 1–5.
14. D. He, C. Liu, T. Q. S. Quek, and H. Wang, "Transmit antenna selection in MIMO wiretap channels: A machine learning approach," *IEEE Wireless Commun. Lett.*, vol. 7, no. 4, pp. 634–637, Aug. 2018.

15. S. Kaviani, O. Simeone, W. A. Krzymien, and S. Shamai, "Linear MMSE precoding and equalization for network MIMO with partial cooperation," in *Proc. IEEE Global Commun. Conf.*, Dec. 2011, pp. 1–6.
16. C. Kuziemsky, "Decision-making in healthcare as a complex adaptive system," *Healthc Manage Forum*, vol. 29, no, 1, pp. 4–7, 2016. https://doi.org/10.1177/0840470415614842.
17. Deloitte. "Global health care outlook: The evolution of smart health care," 2018.
18. D. W. Young and E. Ballarin. "Strategic decision-making in healthcare organizations: It is time to get serious," *Int. J. Health Plann. Manage*, vol. 21, no. 3, pp. 173–191, 2006.
19. J. V. Singh. "Slack, and risk taking in organizational decision-making," *Acad. Manag. Ann.*, vol. 29, no. 3, pp. 562–585, 1986.
20. B. R. A. Cirkovic, A. M. Cvetkovic, S. M. Ninkovic, and N. D. Filipovic, "Prediction models for estimation of survival rate and relapse for breast cancer patients," *IEEE 15th Int. Conf. Bioinform. Bioeng. (BIBE)*, pp. 1–6, 2–4 Nov. 2015.
21. D. Combs and S. Safal Shetty, "Big-data or slim-data: Predictive analytics will rule with world," *J Clin Sleep Med.*, vol. 22, no. 2, 2016.
22. D. Pop and G. Iuhasz, "Overview of machine learning tools and libraries," Tech. Rep. IEAT-TR-2011.
23. A. T. Eshlaghy et al., "Using three machine learning techniques for predicting breast cancer recurrence," *J. Health Med. Inform*, vol. 91, 2013.
24. R. French et al., *Organizational behaviour.* 2nd ed. New York: John Wiley & Sons; 2011.
25. H. A. Haenssle et al., "Man against machine: Diagnostic performance of a deep learning convolutional neural network for dermoscopic melanoma recognition in comparison to 58 dermatologists," *Ann. Oncol.*, pp. 1836–1842, 2018.
26. J.-Z. Cheng et al., "Computer aided diagnosis with deep learning architecture: Applications to breast lesions in us images and pulmonary nodules in CT scans," *Sci. Rep.*, vol. 6, p. 24454, 2016.
27. M. Cicero et al., "Training and validating a deep convolutional neural network for computer-aided detection and classification of abnormalities on frontal chest radiographs," *Invest. Radiol.*, vol. 52, pp. 281–287, 2017.
28. T. Kooi et al., "Large scale deep learning for computer aided detection of mammographic lesions," *Med. Image Anal.*, vol. 35, pp. 303–312, 2017.
29. C. M. Barreira et al., "Abstract WP61: Automated large artery occlusion detection in st roke imaging-paladin study," *Stroke*, vol. 49, p. AWP61, 2018.
30. V. Gulshan et al., "Development and validation of a deep learning algorithm for detection of diabetic retinopathy in retinal fundus photographs," *JAMA*, vol. 316, pp. 2402–2410, 2016.

6 Hardware Trojans in Microfluidic Biochips
Principles and Practice

Dilip Kumar Dalei and Debasis Gountia

CONTENTS

6.1 INTRODUCTION

A hardware Trojan (HT) is defined as a malicious component that is intentionally placed inside an Integrated Circuit (IC) to carry out harmful activity. HTs are basically modeled as a covert tweaking of the original chip circuit to modify the desired functionality. Normally, these get activated in the event of unpredictable inputs or trigger conditions. The activation of a Trojan may stop the normal operation of the chip or create a backdoor to leak secret information. HTs can be differentiated based on their size, complexity, lethality, etc. Researchers have proposed numerous detection techniques to identify HTs in an electronic chip. The form and complexity of these methods solely depend on the type of targeted

DOI: 10.1201/9781003229704-6

HT. Each method has its own merits and demerits, depending on factors such as basic principle, cost, reliability, etc.

Microfluidic biochips are an emerging branch of lab-on-chip (LoC) technology in biochemistry and molecular biology. These biochips help to integrate conventional procedures in biochemical analysis onto a single chip. So, it has a wide global market application in the area of clinical trials, drug therapy, DNA sequencing, etc. A biochip is a miniaturized device similar to an electronic chip that is used to analyze organic molecules associated with living organisms. It is specially designed for functioning in a biological setup, especially inside living organisms. A biochip consists of millions of biosensors, which act as micro-reactors used for detecting particular analytes, such as enzymes, proteins, biological molecules, and antibodies.

The remainder of this chapter is organized as follows. Section 6.2 presents an overview of the background of different types of existing hardware Trojans with their characterization and detection mechanisms. Section 6.3 elaborates on different types of microfluidic biochips with background and working principles. Section 6.4 assesses hardware Trojans in microfluidic biochips along with security and reliability issues, respectively, associated with different microfluidic biochips, including a survey of already published techniques regarding their security concerns. Section 6.5 concludes the chapter with future scope.

6.2 HARDWARE TROJAN

6.2.1 Overview

Hardware Trojans are a rising threat in the area of electronic chips and systems. Due to advancements in IC technology, the chip has become ubiquitous in all electronics systems. Nowadays, chip-based electronics are a central component of all private and government IT systems. The heavy reliance on electronic chips has made these systems more vulnerable to hardware-level security threats. HT is a security threat that surfaces at the lowest level of system security, i.e., at the level of electronics chips and boards. The chip needs to be more trustworthy and reliable when faced with HT security attacks. The detection and identification of security threats at the hardware level are more difficult than at the software level [1].

A HT is a malicious component placed inside a chip to accomplish harmful activity. The Trojans are built by making secret changes in the original design of the chip circuit. The changes in the circuit do not affect the expected functionality; rather, the Trojan gets activated at a later time upon a special condition. Furthermore, the design is also made to avoid detection as much as possible. The target of a HT is either to stop the normal operation of a critical function or to create a back door to leak crucial information.

A HT is always designed with two aspects in mind: malicious intent and how to avoid detection mechanisms. A variety of Trojans have been designed by adversaries to cause different security threats to electronic systems. They differ in factors such as size, complexity, lethality, etc. One such widely adopted nomenclature for Trojans is based on their physical representation and behavior. Their physical representation is concerned with the hardware placement of circuit elements and their inter-connections. Their behavior is characterized by two important quantities: Trigger (Activation stage) and Payload (Action stage) [2]. Below are the key terms related to HTs with their meanings:

- Trigger: an event that initiates the HT. When this particular event starts, the HT circuit is automatically activated for deadly functionality. There are two activation mechanisms that describe the internal and external triggering mechanism of hardware Trojans, i.e., executed for a particular instance of time, and executed externally, respectively.
- Payload: an event that activates the Trojan responsible for implementing HT attacks, which could result in serious effects, such as information leakage, denial-of-service (DoS), and chip reliability degradation.

Hardware Trojans can be inserted into a chip during its fabrication, design, assembly, in-field design, calibration, and testing.

6.2.2 Characterization

The characterization of Trojans depends primarily on two aspects: physical implementation and Trojan behavior. Depending on the physical implementation, a Trojan can be either parametric or functional in nature. Parametric Trojans make changes in the original circuit design itself; they simply modify the intended functionality of a chip. The modification takes place in the internal components of the original circuit, e.g., wire thinning, weakening of flip-flops using radiation, etc. On the other hand, functional Trojans add or delete components in the original design of the chip. They do not alter the basic functionality of the chip. Also, there are other physical aspects that characterize a Trojan, such as size, distribution, and structural alterations.

A Trojan can also be characterized based on its behavior. These behaviors are attributed to how the Trojan is activated and what effects it can bring after activation. There are two basic parts in any Trojan: trigger and payload [3]. A trigger is a module that waits for an event to activate the Trojan. This usually happens during a rare occurrence, so that Trojan can go unnoticed in standard chip testing procedures. Events like undefined input sequences or other mechanisms, like side-channel signals and geographic location, are used for this purpose. Triggers may be combinational or sequential in nature. A combinational trigger waits for a rare combination of one or more input signals. Once the desired inputs come, the trigger gets activated. It again gets deactivated on a change in the signals. On the other hand, a sequential trigger is programmed to remember different states. The states change when rare events occur and trigger the Trojan. **Figure 6.1** illustrates a sequential and combinational circuit demonstrating the presence of HTs [2].

6.2.3 Detection Mechanisms

Numerous detection techniques are proposed to identify a Trojan affected chip. Each approach has its own merits and demerits, depending on factors such as basic

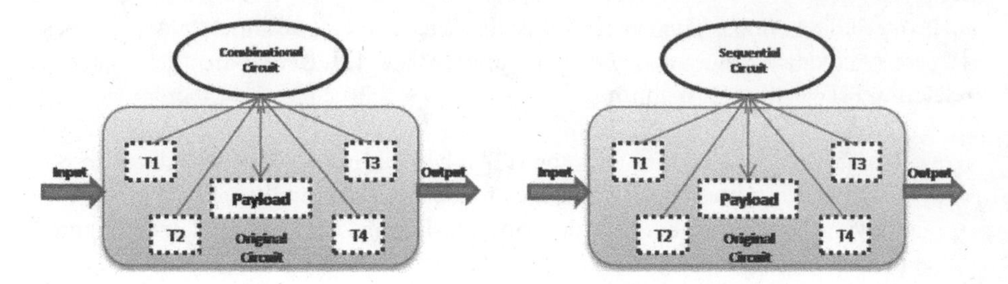

FIGURE 6.1 Structure of combinational Trojan (left) and sequential Trojan (right).

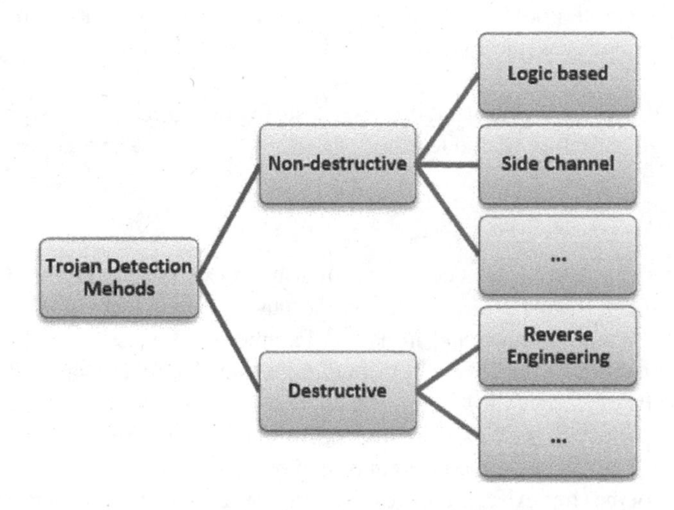

FIGURE 6.2 A simple taxonomy of Trojan detection methods.

principles, costs, reliability, etc. In fact, it is quite hard to designate a single taxonomy that categorizes all kinds of Trojans. But the detection techniques can be classified based on a simple and intuitive taxonomy, as shown in **Figure 6.2**.

The destructive method works by breaking the chip physically for inspection. The physical access to the chip makes the investigation easy and intuitive, but it destroys the chip in the end. The chip is opened layer-by-layer for detailed examination. The chips are physically treated with different chemicals to open the outer layers, and the internal layers are scanned to extract the circuits. There are various visual scanning methods – scanning optical microscopy (SOM), scanning electron microscopy (SEM), pico-second imaging circuit analysis (PICA), voltage contrast imaging (VCI), light-induced voltage alteration (LIVA), and charge-induced voltage alteration (CIVA). Out of these, the most promising technologies are SEM and LIVA. The components of the chip are extracted using reverse-engineering (RE) techniques. But a modern chip is packed with millions and millions of gates, and inspection by RE techniques becomes tedious and time-consuming. The process is also very expensive due to sophisticated and expensive scanning devices. Also, these devices require specialized laboratories with skilled manpower. Of course, the positive side of the technique is that it can be considered as a post-analysis process, which can establish Trojan evidence, once the chip is verified by some other method.

The non-destructive methods detect Trojans inside a chip without physically damaging it. Unlike the destructive method, this method is cost-effective, scalable, and reliable. There are mostly two categories of destructive methods: logic-based and side-channel based. The logic-based analysis is based on the process of finding malicious circuits by functional testing of the IC. It examines the functional behavior of the chip, and any such deviation flags the problem inside the chip. In the test, test vectors are injected into the electronic circuits, and the outputs are observed for any functional anomaly. The specific test patterns are generated through a procedure called automatic test pattern generation (ATPG). This technique detects Trojans by measuring and analyzing the physical parameters of a chip. The parameters are believed to establish a side channel to reveal critical information and hence are also known as a side-channel parameter. The chip has various electronic components that have different physical characteristics. Examples are power consumption, path delay, electromagnetic radiation, thermal profile, etc. These parameters are capable of providing information about the data and state of the device. In this method, the physical parameters of the golden (original) chip are calculated and compared with corresponding parameters of the different fabricated chip of the same design.

6.3 MICROFLUIDIC BIOCHIPS

6.3.1 Background

Microfluidity is the science and technology for controlling extremely small volumes of liquids on the scale of submilliliters or nano-liters. A micro-fluidic biochip integrates various biochemical analysis jobs like point-of-care clinical diagnostics [4] and DNA sequencing [5], etc. on a tiny chip. It facilitates miniaturizing and automating macroscopic biochemical experiments at an extremely small level (10^{-9} or 10^{-18}) [6, 7].

Microfluidic biochips are a growing area of research and development in the field of LoC technology [8]. The major factors that are fueling the biochip technology market are the increasing demand for point-of-care testing, increase in chronic disease incidence, and rising application of proteomics and genomics in cancer therapy. Furthermore, due to the recent outbreak of COVID-19, the demand for lab-on-a-chip and microarray technologies have seen a remarkable rise in the market. The global pandemic has steered the focus on the development of less time-consuming and miniaturized diagnostic kits based on lab-on-a-chip and microarray techniques in pharmaceutical R&D.

6.3.2 Flow-Based Microfluidic Biochip

A flow-based microfluidic biochip, in short FMFB, is the first generation of microfluidic biochip that works on manipulation of a continuous flow in a small volume. The basic building blocks of the FMFB are microvalves, microchannels, and pumps [9].

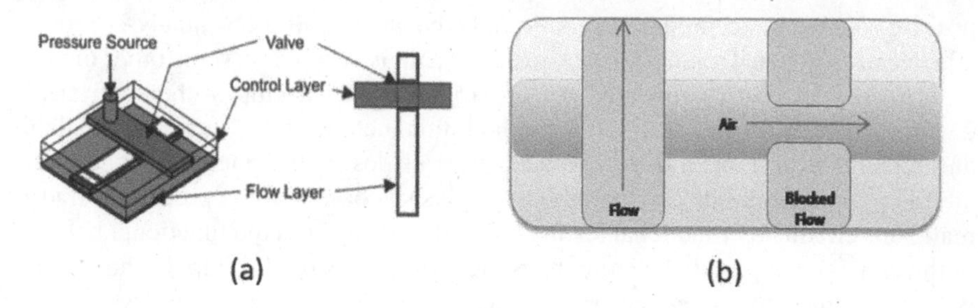

FIGURE 6.3 Schematic diagram of flow-based biochip (a) and control mechanism of the valve (b).

It uses these microcomponents to control liquid flow in the biochip. Where CMOS chips are manufactured on the silicon substrate, FMFB chips are fabricated on an elastomeric material like PolyDiMethylSiloxane (PDMS).

In the FMFB chip, the actuation and movement of fluid flows are controlled through a microfluidic valve. The valves work as gateways that pass or block the flow based on the pressure on the control channel. When pressure is exerted on the control channel, the elastomer will contract the lower layer and block the flow [10]. These microvalves are used to build other components such as mixers, switches, pumps, etc. This is illustrated in **Figure 6.3**.

6.3.3 Droplet-Based Microfluidic Biochip

A droplet-based microfluidic biochip, in short DMFB, is the second generation of microfluidic biochip that works on the manipulation of discrete droplets on a two-dimensional electrode array. The DMFB chip works on the principle of electrowetting-on-dielectric (EWOD) phenomena [11]. In the EWOD phenomenon, a fluid droplet can be manipulated by adjusting the contact angle between a droplet and the substrate using suitable electric potential [12]. The EWOD changes the surface tension upon the application of an electric field. The voltages on the electrodes (DMFB cell) help to control the wetting forces on the droplet. A droplet can be moved to the neighboring cell by keeping a higher voltage in the next cell. This results in a controlled movement of droplets on the chip both in the horizontal and vertical directions [13].

A DMFB chip consists of glass substrates, dielectric layers, hydrophobic layers, continuous ground electrodes on the top plate, and discrete control electrodes on the bottom plate [14]. The chip has two parallel plates – the top plate and the bottom plate. The bottom plate contains addressable electrodes to actuate the fluid droplets, while the top plate is used as a reference electrode. Both the plates are coated with a dielectric layer and a hydrophobic layer. The complete space between the plates is filled with silicone oil medium to avoid evaporation, contamination, and movement of the droplets. The droplets of the fluid samples and reagents are moved and operated within the oil medium. The structure of the biochip is highlighted in **Figure 6.4**.

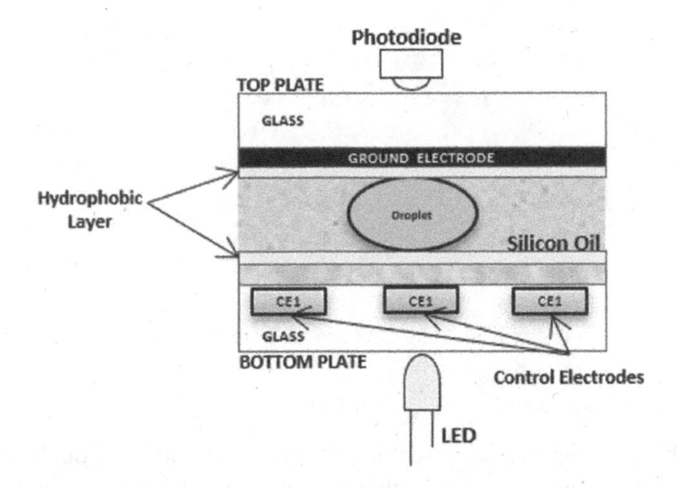

FIGURE 6.4 Schematic of a basic cell for droplet-based microfluidic biochip.

6.4 HARDWARE TROJANS IN MICROFLUIDIC BIOCHIPS

6.4.1 Overview

Biochips are a cheap, portable, reliable, and fast diagnostic tool for many pharmaceutical industries, bio-industries, and medical diagnoses, especially real-time point-of-care diagnostics [15]. The biochip market has become lucrative and has grown exponentially in the past few years. As per a recent report in [16], the current market size of biochips on the global level was USD 15.76 billion in 2021. The value of the market is forecasted to grow at a CAGR of 22% to reach USD 42.96 billion by 2026. This opens up a plethora of security threats from malicious agents in order to achieve either quick financial gains or adversarial gain by manipulating biochips [17]. Like conventional electronic hardware, biochips are also susceptible to security threats such as RE attacks, HTs, IP Theft, etc. One such threat is a HT attack, which has drawn a lot of attention to the prevention and detection of HTs in the electronic chip sector. A HT attack in a biochip basically modifies the internal circuit to affect the test outcome for various malicious purposes, leading to a state of bioterrorism [18]. Trojan attackers insert malicious components inside the chip, which leads to a serious error in the results of microfluidic biochips engaged in different applications.

There are two prime differences in HT attacks between conventional digital circuits and MFB [19]. First, the conventional chip is fabricated using an opaque outer layer, while MFB has transparent outer covers, which allows attackers to view first-hand the structural details of the chip. The transparent properties of MFBs facilitate RE attacks followed by insertion of HTs. Second, conventional chips have a longer life than MFBs, which are generally consumable in nature. The traditional methods for HT detection, such as logic testing and side-channel analysis [2], cannot be directly applied to MFBs. These detection techniques need to execute protocols on the target biochip, which lessens the lifetime of the biochip.

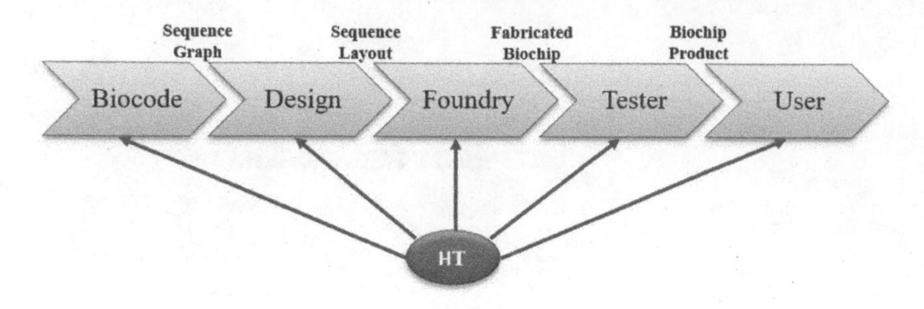

FIGURE 6.5 General supply chain of a biochip and HT threat scenario.

6.4.2 Threat Scenario

Each biochip goes through a well-defined supply chain, starting from specification to manufacturing. There are different phases for different activities related to biochips, such as biochip specification, design, fabrication, testing and calibration, and assembly [20]. Each phase needs appropriate protection from the possibilities of HT. The general supply chain of a biochip is illustrated in **Figure 6.5**.

A. **Specification/Biocoding**

The specification phase, also known as biocoding, represents the high-level specification of an assay. The assay is expressed in the form of a sequence graph along with assay completion time and chip size. A malicious biocoder can intentionally modify the assay operation time to manipulate the outcome.

B. **Design**

The design stage creates the actual sequence and the chip layout from a high-level specification. It uses various kinds of microfluidic functional modules such as mixers and storage units from an existing microfluidic library. A malicious designer can temper the actual sequence graph to modify the operation time leading to incorrect results.

C. **Fabrication**

In the fabrication stage, a biochip is manufactured from the layout supplied from the design stage. The fabrication stage can also be used to implant a Trojan inside a biochip. For example, the actuation voltage across the electrode can be manipulated. Less voltage may actuate the droplet sequence, while an excess of voltage may shorten the lifetime of the chip.

D. **Testing and Calibration**

During testing and calibration, a biochip is measured for any kind of manufacturing defect. A malevolent biochip tester can also embed a Trojan maliciously through embedded capacitive sensors or optical detectors, which may affect runtime readings during the calibration and testing phase.

E. **Assembly**

The assembly process arranges the biochip along with other hardware components on a single printed circuit board (PCB). The components include a ring oscillator circuit, a signal-processing module, a shift-register bank, and a controller memory. These components ensure the robust control of the chip. A malicious assembly may introduce security loopholes by tempering onboard components, e.g., a signal-processing unit may be altered to give the wrong output.

F. **In-Field**

A biochip is configured to execute for a specific bioprotocol. In the field, an attacker can alter this bioprotocol by modifying the actuation sequence. At this stage, a malicious user can insert a software Trojan through network interfaces. It is not possible to introduce a HT at this stage, but a user can exploit a HT already inserted in the chip at earlier stages of production.

6.4.3 Trojan Activation

The activation mechanism of a HT in a biochip defines how the Trojan gets executed upon a trigger. An active Trojan is always running in a biochip or can be triggered by an internal or external event. For example, Trojans designed to make changes in the calibration curve for a glucose bioassay always stay active, whereas those meant to alter the mixing time or the incubation time depend on a trigger [20].

Internal triggers can be designed based on time parameters like clock cycles. For example, the timer can be manipulated for the thermal cycling module. Similarly, fluidic conditions can trigger a Trojan, for example, when a sample concentration goes below a certain threshold.

External triggers can be made on various biochip parameters such as a specific reservoir's fluidic content, a specific detector's output, etc. For example, a primer reservoir can be filled with a light-generating fluid (luciferase) so that when the droplet is dispensed and mixed with a DNA sample, the emitted light intensities (based on a target DNA sequence) are sufficient to trigger a Trojan [20]. The external trigger can be in the form of a malicious actuation sequence that can activate the Trojan.

6.4.4 Trojan Effects

The objectives of a Trojan in a biochip are modified functionality, performance degradation, leaking of secret information, and denial-of-service attacks. The functionality of a specific module can be changed by a HT to produce the wrong assay result. For example, a Trojan can affect the intra-magnet module, causing faulty bead snapping in immunoassay protocols. Similarly, a Trojan can cause excess delivery of voltage which can eventually degrade the electrowetting properties of the electrodes over time. The secret information regarding bioprotocols or actuation sequences can leak, causing IP theft of the biochip. The steps of a protocol can be leaked to deduce the complete bioassay. The bioassay is an intellectual property of a company, which can be stolen to cause great financial loss. In a DoS attack, the correct operation of a biochip can be prevented by a Trojan. For example, a Trojan can intentionally contaminate a droplet by forcing a droplet down the wrong route to affect the residues of another droplet.

6.4.5 Trojan in DMFB

The droplet-based MFB chip is vulnerable to Trojan threats during all stages of the supply chain, as described earlier. Like CMOS chips, a DMFB is also a vulnerable platform for Trojan insertion. A Trojan in DMFB can be used to leak a proprietary bioprotocol or damage a chip to make it unusable for assay operations [21]. Furthermore, Trojan attacks are also applicable in CMOS and can also be applied to DMFB chips.

6.4.6 Trojan in FMFB

A FMFB manipulates the continuous flow inside using microchannels and microvalves. Like digital MFBs, FMFBs also have enough attack surfaces that make them vulnerable to potential HT attacks. There are a lot of common threat surfaces for both DMFB and FMFB chips. HTs can be planted at different phases of the chip, as described earlier.

The structure of a FMFB chip can be broadly grouped into two units: the on-chip control circuit and on-chip fluid handling. The Trojan attack can place a HT either in an *on-chip control circuit unit* or an *on-chip fluid handling unit*. This is illustrated in **Figure 6.6**. The HT placement in the control unit can initiate a signal which can activate the HT, i.e., the output from the control unit will be a trigger for activation of the implanted HT [19].

The sequence graph is a graphical representation of the bioprotocol and is also vulnerable to HT attack. This graph can be modified to implant a HT that can affect the bioprotocol process. For example, an attacker can add an extra mixing step that can contaminate the reagent. It can also activate extra heating operation to affect the fluid sample, resulting in invalidation of the sample. There is a possibility of deleting an operation present in the original protocol.

6.4.7 Defense Mechanisms

There are a lot of defense mechanisms designed to protect biochips from various security threats, such as RE attack, IP piracy, HT, etc. [22]. Some well-known mechanisms are watermarking [23], hardware metering [24], side-channel fingerprinting [1], microfluidic encryption [25], obfuscation [26], randomized checkpoint [27, 28], locking [29], and hamming distance [30], which can be applied

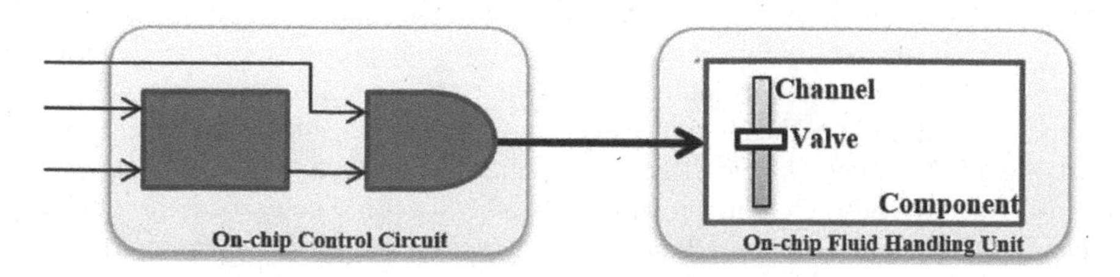

FIGURE 6.6 Structure of on-chip control and fluid handling unit in FMFB chip [19].

to biochips to defend against HT attack. Except for watermarking and hardware metering, the rest are capable of protecting a biochip against HT. Watermarking protects the ownership of a biochip in cases of IP theft by embedding a signature in the chip design. Similarly, hardware metering adds the buyer's signature to the biochip in addition to the owner's signature (as in watermarking) to ensure IP legitimacy and the source of leakage.

A. **Microfluidic Encryption**

Microfluidic encryption can prevent the insertion of malicious Trojans in a biochip [25]. A key is used to encrypt and decrypt the bioassay protocol. Only an authenticated person with the correct key pattern can carry out the synthesis of biochemistry protocols. The microfluidic level encryption is especially effective to produce the correct bioassay output with integrity. In this case, a sequencing graph G is encrypted to form another sequencing graph G' through a series of secret keys. The keys are only known to an authorized user. During assay execution, the correct key enables the bioassay and allows the desired sequence of microfluidic operations. On the other hand, if the user-supplied key mismatches with the secret key, the flow of droplets through the DMFB will be blocked, and no results will be produced by the system. For example, [25] has proposed a microfluidic encryption methodology which consists of the following:

a. Two input fluidic data,
b. One fluidic control input, i.e., one bit of the secret key,
c. One fluidic data output.

A number of such microfluidic multiplexers are added in the design flow to form G' from G. The secret key consists of logic 1, which means associated presence, and logic 0, which means the absence of control droplets. Only an authorized user can forward those required droplets by opening all the fluidic multiplexers. One security enhancement of the proposed method is the aging factor of a chip. Aging enhances the security of the DMFB chip as it allows fewer trials to get the security key to break the scheme. This is due to the fact that biochip degradation starts quickly and should be discarded within a few hours [31], and electrodes can be actuated for a limited time before dielectric breakdown occurs [32]. One of the security benefits of the fluidic multiplexer is protection against HT attacks. The HT manipulates the assay result by altering the sequencing graph. The attacker must need prior knowledge of the assay to launch these attacks. Microfluidic encryption obfuscates the assay so that the assay cannot be manipulated to get a meaningful outcome that can pass scrutiny. The other security benefit is DMFB chip supply-chain security, where overproduction and counterfeiting of DMFB chips are meaningless without the knowledge of the secret key. Also, area overhead is defined as the number of electrodes per electrode array [33]. The number of electrodes is directly proportional to the number of multiplexers. For this reason, the number of multiplexers should be placed in the correct places.

B. **Randomized Checkpoint**

The randomized checkpoint method can be used to detect HTs in a biochip. It takes a random measurement of the location and time of the biochip with an on-chip charge-coupled device (CCD) camera [34]. In other words, it audits the progress of a biochemical assay on-chip at particular locations and times using various sensors found in the error re-recovery system. There are also processing and memory constraints for the method. This method also describes the proper evaluation of a sample. In the randomized checkpoint method, uncertainty for the attacker increases. Also, random sampling electrodes for a checkpoint operation brings down the cost, but still provides a finite guarantee of security. A randomized checkpoint-based error recovery system is capable of defending against these attacks and also has the advantage of being lightweight and easy for biochip researchers and engineers to implement in patho laboratories, biochemical and biological industries.

In the randomized checkpoint proposed in [34], a secure coprocessor was designed to monitor the execution of assay operation on the chip. The coprocessor is placed with the CCD camera found in many cyber-physical DMFBs for checkpoint-based error recovery. The advantage of the coprocessor is that it is not prone to any network attacks due to the physical separation. The original copy of the golden bioassay specification, which is secured during manufacturing, is kept in the processor. The system operation of the proposed method can be described in the following steps.

- Random selection of electrodes at the current time step.
- Random selection of an electrode for inspection.
- Extraction of electrode state.
- Comparison with golden actuation sequence.
- Selection of next electrode.
- Repeat for each time step.

C. **Obfuscation**

The obfuscation method protects a biochip from a RE attack, which can lead to the implant of a HT in the biochip. It obfuscates the architecture and actuation sequence of the chip. It indirectly protects a biochip from a Trojan attack. In this method, the functionality of the design is hidden by inserting additional elements, and only correct inputs to the added elements are essential to exhibit the desired functionality.

The code-obfuscation technique may be used by a DMFB chip designer for the mystification of actuation sequences; hence, this obfuscation is able to protect against Trojan attacks indirectly as an attacker may not be able to find meaningful and stealthy hardware Trojans with such an obfuscated sequence. Obfuscation is able to prevent Trojans and also reverse engineering, but not piracy and counterfeiting.

D. **Side-Channel Fingerprinting**

Side-channel fingerprinting is a popular method for detecting Trojans by observing changes in chip parameters like area, power, delay, etc. These parameters of a manufactured biochip are compared with the gold biochip for detection of the Trojan. This technique is able to detect hardware Trojans easily as the manufactured parametric characteristics, such as power, area, delay, and droplet characteristics of the chip, are compared with those of the statistical model. Any significant variation/deviation would be considered a Trojan. Side-channel fingerprinting is not able to prevent piracy, reverse engineering, and counterfeiting attacks.

E. **Locking**

The locking mechanism protects a biochip from the insertion of a HT during manufacturing in a factory. It puts a lock mechanism in tamper-proof memory on the chip. The protocol can be executed only with the correct key. Any operation without a correct key will run the protocol in the wrong order and produce a wrong result.

A DMFB chip designer can add locks (i.e., fluidic multiplexers) that manage and control the droplet flow between electrodes or other microfluidic components. These droplets will flow correctly only if the correct key is applied; otherwise, the wrong droplets will flow, which leads to a wrong output. This key should be stored in tamper-proof memory in order to prevent vulnerabilities as the key is erased during reverse engineering. Trojans cannot be inserted as the DMFB chip functionality is locked/hidden by the key. Locking prevents all four attacks: Trojans after fabrication, piracy, reverse engineering, and counterfeiting attacks, except for Trojans inserted during biochip fabrication in the factory.

F. **Code Analysis**

Code of actuation sequences are analyzed to detect any Trojans inserted in the system. Also, any advanced encryption algorithms and hash functions can be used for the confidentiality of actuation sequences and hence prevent Trojan attacks. Code analysis is not able to protect the biochip against piracy, reverse engineering, and counterfeiting attacks.

G. **Hamming Distance**

Hamming distance is a method proposed to detect Trojans that can modify bioprotocols by changing the actuation sequences in the field [30]. This model can effectively discover any arbitrary changes in the original actuation sequence. Hamming distance is a simple and effective technique to detect and verify errors in a sequence [35]. The same concept can be applied to compare the tampered actuation sequence and the golden activation or actuation sequence.

H. **Metering**

In this technique, both the buyer's public signature and the designer's digital signature are added to the chip design as a synthesis constraint. This metering technique is also not able to provide protection against hardware Trojans as the attacker is able to create and hide a Trojan in the biochip due to the availability of the chip design functionality.

I. **Watermarking**

In this technique, the digital signature of the original designer is added to the biochip design. Watermarking is able to prove ownership as these

TABLE 6.1
Summary of Potential Defenses in Microfluidic Biochips

	Name of Attacks			
Name of Defense	**Trojans**	**Piracy**	**Reverse engineering**	**Counterfeiting**
Hamming distance	No*	Yes	No	Yes
Watermarking	No	No*	Yes	No*
Metering	No	No*	Yes	No*
Side-channel fingerprinting	No*	No	No	No
Code analysis	No*	No	No	No
Obfuscation	Yes	No	Yes	No
Locking	Yes*	Yes	Yes	Yes
Microfluidic encryption	Yes*	Yes	Yes	Yes
Randomized Checkpoint	Yes*	No**	No	No*

are very difficult to identify and alter. This watermark is embedded as a synthesis constraint by the DMFB chip designer during DMFB chip design/actuation sequence generation. But watermarking cannot provide security against Trojans.

The potential defense mechanisms are summarized in **Table 6.1**. Depending on their business strategy and budget, DMFB chip designers can choose any one or multiple aforementioned techniques to protect the biochip against HTs. The symbols in **Table 6.1** mean:

- **Yes** means both detection and prevention are possible.
- **Yes*** means to detect and prevent those Trojans inserted only after fabrication, but not those before fabrication.
- **No** means no detection and no prevention.
- **No*** means only detection, but no prevention.
- **No**** means only prevention, but no detection.

6.5 CONCLUSION AND FUTURE SCOPE

Biochips are a lab-on-chip technology that helps to build miniature automated health analysis systems for lab-level manual biochemical processes, such as biochemical testing, DNA analysis, immunoassays, environment monitoring, clinical diagnostics, point-of-care disease treatment, etc. The market value of biochips is expected to rise to the tune of USD 42.96 billion by 2026. The growing market for biochips is mainly due to the rapid occurrence of chronic disease, increased cancer treatment and research, and the onset of frequent global pandemics. The current COVID-19 pandemic has also given a further impetus to the demand for biochips.

This chapter has presented a horizontal study of security threats and countermeasures in the form of hardware Trojans in biochip technology, especially DMFB and FMFB biochips. The fundamental design and structure of a HT in an electronic chip have also been explained. Like a CMOS chip, a biochip is also vulnerable to HT attacks and needs considerable attention for reliable and efficient protection. In fact, most CMOS chip related HT threats are equally applicable to the biochip. Of course, HT attacks in biochips are considered easier as compared to other ICs due to the reasons mentioned in the chapter. There are multiple and equal threat preceptors available in both types of biochip – FMFB and DMFB at various stages of their supply chain.

There is a still a lot of challenges in the search for a secure and robust mechanism for protecting biochips from HT attacks. Efficient encryption techniques are being built for sequence graphs and layouts in biochips. In the future, an efficient key management scheme can be designed to allow a unique key for each actuation block. Also, it will be an upcoming research area to devise a spectroscopy resilient locking mechanism for bioprotocols from HT threats. There is also a need for better Boolean functions to compute efficient tweaks to protect IP theft from cryptanalytic attacks, which indirectly encourages the implantation of HTs. Currently, there is no single method to secure microfluidic biochips against rising security attacks such as hardware Trojans, denial-of-service, IP theft, and so on. Thus, there is a need to establish a secure and reliable mechanism that will make microfluidic biochips trustworthy to avoid attacks on the world of healthcare, biological, and biochemical industries for a sustainable biochip world.

REFERENCES

1. M. Tehranipoor, and F. Koushanfar, "A survey of hardware Trojan taxonomy and detection." *IEEE Design and Test of Computers*, vol. 27, no. 1, pp. 10–25, 2010.
2. S. Bhunia, M. S. Hsiao, M. Banga, and S. Narasimhan, "Hardware Trojan attacks: Threat analysis and countermeasures." *Proceedings of the IEEE*, vol. 102, no. 8, pp. 1229–1247, 2014.
3. J. Rajendran, E. Gavas, J. Jimenez, J. Padman, V. Padman, and R. Karri, "Towards a comprehensive and systematic classification of hardware Trojans." In *IEEE International Symposium on Circuits and Systems (ISCAS 2010)*, pp. 1871–1874. IEEE, 2010.
4. D. Gountia, and S. Roy, "Design-for-trust techniques for digital microfluidic biochip layout with error control mechanism." *IEEE/ACM Transactions on Computational Biology and Bioinformatics (TCBB)*, vol. X(XX), pp. 1–13, 2021.
5. S. Subidh Ali, M. Ibrahim, O. Sinanoglu, K. Chakrabarty, and R. Karri, "Security assessment of cyberphysical digital microfluidic biochips." In *IEEE/ACM Transactions on Computational Biology and Bioinformatics (TCBB)*, vol. 13, no. 3, pp. 445–458, 2016.
6. R. B. Fair et al., "Chemical and biological applications of digital microfluidic devices." *IEEE Design & Test of Computers*, vol. 24, no. 1, pp. 10–24, 2007.
7. M. Yasin et al., "On improving the security of logic locking." *IEEE Transactions on Computer-Aided Design of Integrated Circuits and Systems (TCAD)*, vol. 35, no. 9, pp. 1411–1424, 2010.

8. R. W. Hamming, "Error detecting and error correcting codes." *The Bell System Technical Journal*, vol. 29, no. 2, pp. 147–160, 1950.
9. C. Dong, T. Chen, J. Gao, Y. Jia, P. Mak, M. Vai, and R. P. Martins, "On the droplet velocity and electrode lifetime of digital microfluidics: Voltage actuation techniques and comparison." *Microfluidics and Nanofluidics*, vol. 18, no. 4, pp. 673–683, 2015.
10. D. J. Boles et al., "Droplet-based pyrosequencing using digital microfluidics." *Analytical Chemistry*, vol. 83, no. 22, pp. 8439–8447, 2011.
11. J. Tang, M. Ibrahim, and K. Chakrabarty, "Randomized checkpoints: A practical defense for cyberphysical microfluidic systems." In *IEEE Design Test*, pp. 1–8, 2018.
12. G. M. Whitesides, "The origins and the future of microfluidics." *Nature*, vol. 442, no. 7101, pp. 368–373, 2006.
13. K. Chakrabarty, and F. Su, *Digital microfluidic biochips: Synthesis, testing and reconfiguration techniques.* CRC Press, 2007.
14. J. Tang, M. Ibrahim, K. Chakrabarty, and R. Karri, "Security implications of cyberphysical flow-based microfluidic biochips." In *Proceedings of the IEEE Asian Test Symposium*, pp. 115–120, 2017.
15. F. Su, and K. Chakrabarty, "High-level synthesis of digital microfluidic biochips." *Journal on Emerging Technologies in Computing Systems (JETC)*, vol. 3, no. 4, p. 32, 2008.
16. S. S. Ali, M. Ibrahim, J. Rajendran, O. Sinanoglu, and K. Chakrabarty, "Supply-chain security of digital microfluidic biochips." *Computer*, vol. 49, no. 8, pp. 36–43, 2016.
17. Z. Guo, J. Di, M. M. Tehranipoor, and D. Forte, "Obfuscation-based protection framework against printed circuit boards unauthorized operation and reverse engineering." *ACM Transactions on Design Automation of Electronic Systems*, vol. 22, no. 3, pp. 1–31, 2017.
18. H. Chen, S. Potluri, and F. Koushanfar, "Security of microfluidic biochip: Practical attacks and countermeasures." *ACM Transactions on Design Automation of Electronic Systems*, vol. 25, no. 3, p. 29, 2020.
19. D. Gountia, and S. Roy, "Checkpoints assignment on cyber-physical digital microfluidic biochips for early detection of hardware Trojans." In *Proceedings of Third International Conference on Trends in Electronics and Informatics (ICOEI 2019)*, pp. 16–21, 2019.
20. C. Dong et al., "A survey of DMFBs security: State-of-the-art attack and defense." In *Proceedings of the 21st International Symposium on Quality Electronic Design (ISQED)*, pp. 14–20, 2020.
21. M. Shayan, S. Bhattacharjee, A. Orozaliev, Y. A. Song, K. Chakrabarty, and R. Karri, "Thwarting bio-IP theft through dummy-valve-based obfuscation." *IEEE Transactions on Information Forensics and Security (TIFS)*, vol. 16, pp. 2076–2089, 2021.
22. M. G. Pollack, R. B. Fair, and A. D. Shenderov, "Electrowetting-based actuation of liquid droplets for microfluidic applications." *Applied Physics Letters*, vol. 77, no. 11, pp. 1725–1726, 2000.
23. H. Chen, S. Potluri, and F. Koushanfar, "BioChipWork: Reverse engineering of microfluidic biochips." In *IEEE International Conference on Computer Design (ICCD)*, pp. 9–16, 2017.
24. S. Venkatesh, and Z. A. Memish, "Bioterrorism: A new challenge for public health." *International Journal of Antimicrobial Agents*, vol. 21, pp. 200–206, 2003.
25. M. Shayan, S. Bhattacharjee, J. Tang, K. Chakrabarty, and R. Karri, "Bio-protocol watermarking on digital microfluidic biochips." *IEEE Transactions on Information Forensics and Security (TIFS)*, vol. 14, no. 11, pp. 2901–2915, 2019.
26. F. Koushanfar, "Hardware metering: A survey." In *Introduction to hardware security and trust*, M. Tehranipoor, and C. Wang, eds., Springer, pp. 103–122, 2011.

27. "Global biochips market size, share, trends, COVID-19 impact & growth analysis report – Segmented by type, end user and region – Industry forecast (2021 to 2026)." https://www.marketdataforecast.com/market-reports/bio-chip-market, April 2021.
28. R. Sista et al., "Development of a digital microfluidic platform for point of care testing." *Lab on a Chip*, vol. 8, no. 12, pp. 2091–2104, 2008.
29. S. S. Ali et al., "Microfluidic encryption of on-chip biochemical assays." In *Proceedings of the BioCAS*, pp. 152–155, 2016.
30. F. Mugele, and J. Baret, "Electrowetting: From basics to applications." *Journal of Physics: Condensed Matter*, vol. 17, no. 28, p. R705, 2005.
31. D. Gountia, and S. Roy, "Security model for protecting intellectual property of state-of-the-art microfluidic biochips." *Elsevier Journal of Information Security and Applications*, vol. 58, pp. 1–15, 2021.
32. J. Tang, M. Ibrahim, and K. Chakrabarty, "Randomized checkpoints: A practical defense for cyberphysical microfluidic systems." In *IEEE DesignTest*, pp. 1–8, 2018.
33. M. Shayan, S. Bhattacharjee, R. Wille, K. Chakrabarty, and R. Karri, "How secure are checkpoint-based defenses in digital microfluidic biochips?" *IEEE Transactions on Computer-Aided Design of Integrated Circuits and Systems (TCAD)*, vol. 40, no. 1, pp. 143–156, 2021.
34. J. L. He et al., "Digital microfluidics for manipulation and analysis of a single cell." *International Journal of Molecular Sciences*, vol.16, no. 9, pp. 22319–22332, 2015.
35. H. He, and H. Hu, "Field-level digital microfluidic biochips Trojan detection based on hamming distance." In *Proceedings of the IEEE 4th Information Technology, Networking, Electronic and Automation Control Conference (ITNEC)*, pp. 640–643, 2020.

7 Benefits and Risks of Cloud Computing

Osheen Oberoi, Sahil Raj, Viput Ongsakul, and Vishal Goyal

CONTENTS

DOI: 10.1201/9781003229704-7

7.1 CLOUD COMPUTING

The stupendous breakthroughs in science and technology that the world has witnessed have brought about several blessings to human beings in recent times. One of the blessings of technology to mankind in the era of the Information Age is the innovation of cloud computing.

The prime objective of this modern paradigm is to endow quality service that is reliable and can be tailored as per the needs of consumers [1]. The word "cloud" in the term "cloud computing" is a metaphor that is used for the internet; thus, the phrase cloud computing signifies internet-based computing. The National Institute of Standards and Technology's (NIST) defines the critical aspects of the cloud as

> cloud computing is a model for enabling ubiquitous, convenient, on-demand network access to a shared pool of configurable computing resources (for example, networks, servers, storage, applications, and services) that can be rapidly provisioned and released with minimal management effort or service provider interaction. [2]

Cloud computing enables a set of shared computing services, inclusive of servers, storage, software, analytics, networking, and databases, over the web, i.e., the cloud on a utility basis [3]. The notion of cloud computing is to change traditional desktop computing to service-oriented computing. Cloud technology is popular among individuals and various business organizations as it is a cost-effective model and proliferates the productivity, efficiency, and economies of scale of a business. Along with this, cloud solutions are extremely reliable and also offer disaster recovery. Moreover, this model enables the users with a performance monitoring facility including the hallmark feature of cloud computing to scale dynamically. Undeniably, cloud computing plays an indispensable role in rendering the burgeoning storage and infrastructure demand in the IT field. Besides, cloud service providers authorize users to store and access data, applications, and programs from the cloud server through the internet [4]. Cloud computing models are scalable; therefore they serve the fluctuating needs of organizations and provide them with optimal resource availability at all times.

Most importantly, users can access services at any time and from any location. Companies using cloud solutions do not have to worry about maintaining complex hardware as the cloud providers themselves manage it. Cloud services have irrefutably become the de facto choice of almost all organizations in the contemporary era, irrespective of their size, type, or industry, because of the trend to transform digitally in order to work remotely from any location. In modern times cloud computing is finding a lot of applications in the various functional areas of organizations.

7.2 DESIRED FEATURES OF CLOUD COMPUTING

Cloud computing is an efficacious way to deliver cloud services as it eradicates the need to store and manage one's data. The pay-as-per-use cloud model comprises a spectrum of services that are delivered over the internet. The cloud computing model includes five essential key competencies that are highlighted in the ensuing paragraphs (see Figure 7.1).

7.2.1 On-Demand Self-Service

This is the most vital feature of cloud computing technology. This feature enables the user to use cloud computing resources as per their need, thus eliminating human interaction between the user and the provider [5]. Cloud computing services are

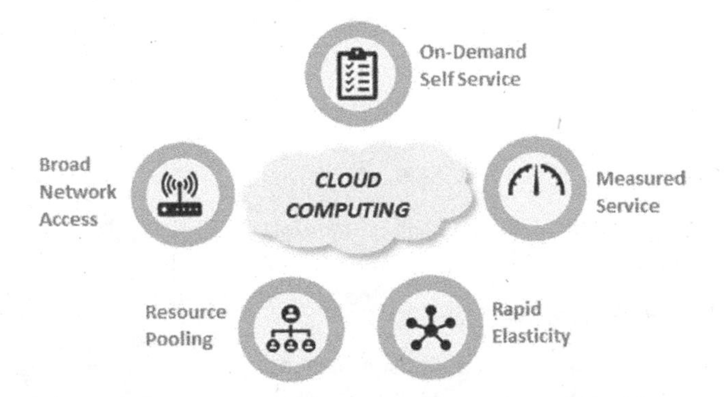

FIGURE 7.1 Desired features of cloud computing. Source: Author.

cost-effective as consumers do not have to invest in buying expensive infrastructures. Moreover, the user subscribes according to their needs [6].

7.2.2 Broad Network Access

The services provided by cloud computing are available over the internet; hence, high bandwidth communication links must be available to connect to the cloud services [7]. As cloud computing is a web-based service, the consumer can access the data at any time and from any distinct location through different devices [8].

7.2.3 Resource Pooling

The cloud vendors utilize a multi-tenant model. In this model, the cloud provider's computing resources are pooled to serve multiple consumers [9]. This means by utilizing the multi-tenancy feature, the cloud providers ensure that privacy and security over the user's information is retained. Thus, resource pooling means that resources are allocated dynamically depending on the requirement of the customer [7, 10].

7.2.4 Measured Service

The user is monitored by the cloud vendors for the number of cloud resources used by them. The user is then automatically billed for using that particular session. Thus, the service providers use appropriate mechanisms to monitor, control, and bill every individual consumer for the usage of the resources through its metering capabilities [11].

7.2.5 Rapid Elasticity

Elasticity is another significant aspect of cloud computing. This feature allows the consumers to rapidly scale the cloud resources either up or down as per the need and demands of the consumer [12]. Moreover, this characteristic is quite cost-effective as it evades the unnecessary wastage of resources. In some cases, it gets scaled automatically according to the consumer's demand, and the services are at the user's disposal 24/7.

7.3 SERVICE MODELS OF CLOUD COMPUTING

Cloud computing has the following service models (see Figure 7.2).

7.3.1 Software as a Service (SaaS)

SaaS is also referred to as cloud application services [13]. This is the most widely utilized service model in the cloud market as it delivers all the applications to its users over the internet. SaaS is a network-based access model, so a third-party vendor

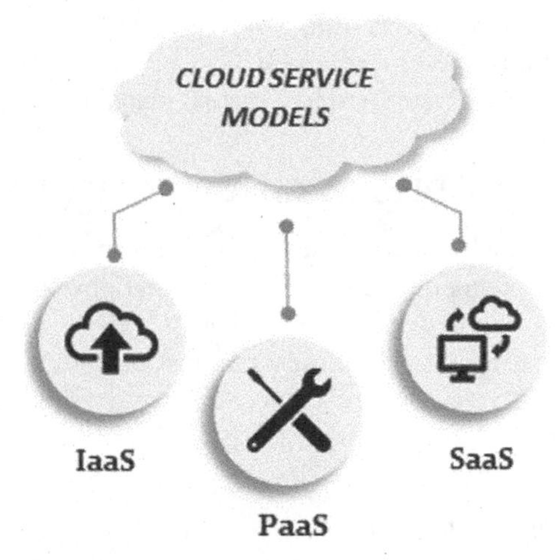

FIGURE 7.2 Service models of cloud computing. Source: Author.

manages it. The user can access SaaS applications either through a web browser known as a thin client interface or through a program interface [14]. The SaaS model curtails the customer's burden of installing, managing, and upgrading the software to a considerable extent [15].

7.3.2 Platform-as-a-Service (PaaS)

This model provides a platform to the consumers. In this model, the clients can either deploy applications created by the consumer or acquire applications created by utilizing different programming languages and libraries that the cloud service provider supports [9]. To a certain extent, this model is identical to the SaaS model, but instead of delivering software over the internet PaaS provides a platform to its users where software can be developed, tested, and deployed [16]. Therefore, PaaS eradicates the user's need to install and buy the required hardware and software and can thus focus on building the software without the hassle of in-house installation. This signifies that solely the client controls the applications deployed and their possible configuration settings [10]. This service model is cost-effective as it is a pay-per-use service that is highly secure, reliable, and can be scaled instantly.

7.3.3 Infrastructure as a Service (IaaS)

IaaS delivers the required infrastructure to its clients, including storage, processing, servers, network resources, and operating systems via virtualization technology [17]. The client can deploy and execute arbitrary software by using this service model. The user does not have to purchase the required servers, data center, or network resources; rather, the user has to pay for the duration they have used the service. The service provider provides virtual servers with unique IP addresses [18]. The users availing of the IaaS service are given complete control over the whole infrastructure.

Clients use the application programming interface (API) to control the servers. Amazon Web Services is one of the largest IaaS provider companies that provide Elastic Compute Cloud (EC2) services like virtual machines for processing [19].

7.4 DEPLOYMENT MODELS OF CLOUD COMPUTING

In this current times, which has witnessed many leaps in science and technological breakthroughs, cloud computing has majorly revolutionized the IT sector. It is essential to determine the deployment model meticulously as each model fulfills different organizational needs; hence, an organization should choose the suitable model. NIST defines four models of cloud deployment, which are elucidated as follows (see Figure 7.3).

7.4.1 Public Cloud

This cloud deployment model runs on the premises of the cloud-like Google and Amazon, which offers services to the consumers and companies through the internet [20]. By using the public cloud, users can access the cloud infrastructure through a web browser. It is owned and managed by cloud service providers [21]. The end users only pay for the time span that they have utilized the cloud services, otherwise known as a pay-per-use basis. Hence, this model is highly cost-effective as it curtails the cost of IT expenditure, such as hardware and application infrastructure. However, the public cloud model is less secure than other cloud models and is not suitable for organizations operating with sensitive information as it is more prone to malicious attacks [14]. Thus, appropriate security checks should be executed by both the organization/client as well the cloud vendor.

7.4.2 Private Cloud

This type of deployment model permits the utilization of cloud resources for a particular organization. Organizations using the private cloud deployment model

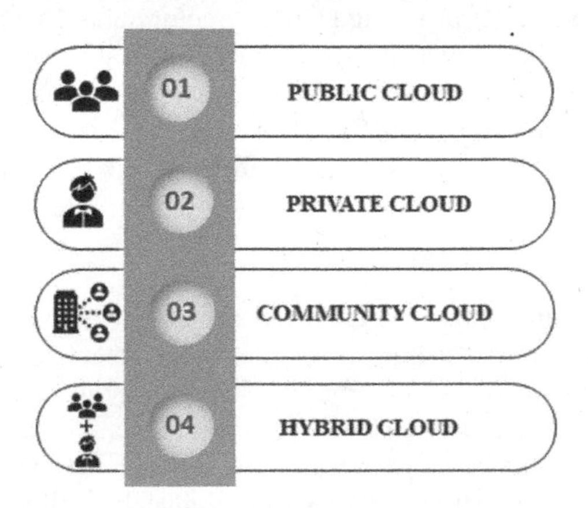

FIGURE 7.3 Deployment models of cloud computing. Source: Author.

can manage this deployment model either independently or a third-party cloud vendor can handle it [22]. It has more control over its resources and hardware as it is fully owned by a single organization [19]. The resources are not shared with other organizations, unlike the public cloud. The organization using this model is solely responsible for its management, maintenance, and updates. Security in the private cloud model is much more enhanced as only users of a single organization have access to this cloud.

7.4.3 Hybrid Cloud

This cloud model is an amalgamation of two cloud services, i.e., the public cloud and the private cloud [23]. The most noticeable feature of this model is that it bounds together with the merits of both types of cloud in an efficient manner. Therefore, a hybrid cloud is an ideal way to procure the benefits as per their need. Here, the consumers place non-business-critical information or less sensitive data on the public cloud. In contrast, the essential core services or business-critical services and data are placed on the private cloud [10]. The different cloud models in the hybrid model are associated with some standard technology; still, they remain a unique entity in the hybrid cloud model [12]. This model also provides the users with a "cloud-bursting" option, which means that if on the private cloud the demand for an application or a resource proliferates; in this case, the company can move to the public cloud [5]. This acts as a huge benefit to the organization as they can then obtain the benefits of additional computing resources and higher capacity.

7.4.4 Community Cloud

This model permits services and infrastructure to be accessible by a collection of organizations with common interests, such as with specific security needs [24]. The maintenance cost is shared among the particular organizations that reap the benefits of this model. Hence, the community cloud model is more expensive as the cost is spread among fewer users than the public cloud [7]. This model aims to accomplish business-related objectives, so this model is apt for those organizations that collaborate on research or joint ventures requiring centralized cloud computing services to manage, build, and implement identical projects.

7.5 ARCHITECTURE OF CLOUD COMPUTING

In the technology-driven world, cloud computing has become a buzzword in the market. This trend in technology is skyrocketing as it is used in all organizations irrespective of their size or industry. Organizations in modern times require huge storage to store their entire data; therefore, these enterprises are in dire need of a technology that can efficiently store heaps of data. Hence, the cloud is the best platform to fulfill this requirement. Thus, organizations can store all their significant information in the cloud to access it anytime and from any location via an internet connection.

Cloud computing architecture has two segments. The first is the front-end segment. The client uses the front-end of the architecture. This implies that this section of the cloud architecture comprises of the user interfaces and the applications. The consumer can access all the cloud resources or services offered by cloud computing by using these interfaces. Second, the back-end section, on the other hand, is utilized by the respective service providers. It comprises numerous servers, data storage systems, and virtual machines, which collectively account for the services facilitated by cloud computing. This segment monitors and manages all the programs that execute applications on the user interface. Moreover, the back-end is also accountable for the endowment of the security mechanism. With the help of middleware, networked computers are allowed to communicate with each other [14].

7.6 BENEFITS OF CLOUD COMPUTING

Cloud computing provides the following benefits to the organizations.

7.6.1 Cost Reduction

This is one of the most lucrative features that clients experience once they move their operations to the cloud environment. The adoption of a cloud infrastructure dramatically reduces infrastructure and operational costs [22]. Consumers no longer have to spend a hefty amount of money on procuring and maintaining computing resources. Also, it rules out the need to hire additional manpower to maintain these systems as the cloud vendors are accountable for both the smooth functioning and maintenance of all the services provided.

7.6.2 Uninterrupted Services

The cloud environment provides uninterrupted services to its clients as the likelihood of outages on cloud systems is relatively minimal. Although some companies using cloud services have encountered outages and disruption to the services in the past, cloud computing resources and services are much more reliable and trustworthy than on-premises services [14].

7.6.3 Easy Implementation

Cloud technology can be purchased and implemented easily in organizations. It only takes a fraction of seconds to set up the cloud system and transfer all the information onto the cloud.

7.6.4 Scalability

Cloud computing benefits organizations by providing flexible and agile platforms that facilitate companies with scalability services. The consumer can scale the required resources up and down anytime as per their need [25].

7.6.5 Green Computing

Cloud computing aims to "go green" and save the costs of enterprises. The computing systems used by organizations produce harmful emissions on a large scale and also generate e-waste, which is highly disadvantageous [3]. This can be reduced considerably by utilizing cloud computing services. This technology helps by curtailing the power, cooling, storage, and usage of space, resulting in sustainable and environmentally friendly data centers. Hence, safeguarding the environment [26].

7.7 CHALLENGES IN CLOUD COMPUTING

Cloud computing is one such technology that is gaining popularity in the IT industry at an alarming rate. Undeniably, it is sprawling at an incredible pace to different organizations across the globe. However, the prime challenges and issues that one encounters with the usage of this technology can also not be overlooked. Confidentiality, privacy, and security are the three prime issues that hinder procuring the numerous advantages endowed by cloud computing technology [26]. As the cloud environment offers a multi-tenancy feature wherein various clients and computers access the same cloud resources or services, security is a big concern [27]. The loss of data and its misuse is one of the key issues consumers face, as malicious hackers can easily breach the data stored on the cloud infrastructure. Through various illegitimate techniques, the hackers acquire illegal access to the data, consequently misusing the owner's sensitive and confidential information. Moreover, the data is stored in remote locations and therefore there is a possibility that the user might be unable to delete their data from the cloud. Another area of concern is related to the security of the data [28]. Anyone using cloud services is potentially at risk of cyber-attack. This usually occurs when attackers use botnet attacks to infect the cloud with malware and spam. Yet another type of security concern is service hijacking, wherein hackers use for several to get unauthorized entry to certain authorized services [29].

7.8 REAL-TIME EXAMPLE: PITFALLS IN AMAZON'S CLOUD SECURITY SYSTEM

Amazon.com, Inc., a company on the frontline that provides efficient cloud computing services, was targeted by cybercrooks. In 2010, the hackers injected a cross-site scripting attack on the cloud infrastructure of Amazon to acquire unauthorized access to its credentials. The adversaries then intruded on the relational database service (RDS) of Amazon. The hackers aimed to get back-end access into the system of Amazon even if they lost the original access they acquired by illegal means. With this attack, the attackers were able to record the login details of any user who attempted to click on the login or sign in button on the homepage of Amazon. The hackers used their servers to disrupt the new machines of Amazon by infecting them with Zeus Trojan horse malware. The attacker gets access to people's email and social media accounts and can steal sensitive information through this attack.

They send messages to the user that appear to be legal. As a result, the computers that were hacked started reporting to Amazon's EC2 for updates and further instructions [30]. Hence, it is evident that security issues in cloud computing can risk the entire network.

7.9 SECURITY ISSUES AND CHALLENGES IN CLOUD COMPUTING

7.9.1 Data Storage and Computing Security Issues

Data in the cloud computing paradigm is considered to be the most imperative asset. The data in the cloud is stored at an isolated location which is hidden from the customers. This is the prime reason that makes the customers unwilling and hesitant to bestow their information on the cloud vendors. Moreover, they are constantly apprehensive of the loss of data into the wrong hands [31], resulting in negative ramifications that might take place at the time of operations and processing. Therefore, while computing, the customers' data must be steady and highly confidential at all processing phases and constantly stored to update the records. There arise a plethora of problems because of this storage. Fundamentally, when the data is stored at a third-party location, the user is completely unaware of what happens to their data after its storage in the cloud [32].

7.9.1.1 Data Storage

Loss of control over one's own data is the foremost issue when storing the data in the cloud [31]. Full control of the data is not provided by the cloud vendor, which is a cause of concern among the users. Moreover, it is a difficult and laborious task to verify the integrity and confidentiality of the data while it is deployed on the cloud. The cloud service providers store the cloud data in a pool known as a server pool, whose location is unknown to the data owner and is controlled and administered solely by the cloud service providers. Only a restricted level of control is permitted to the user, and that too on virtual machines. The two essential characteristics of cloud computing, i.e., virtualization and multi-tenancy, make the data more prone to malicious attacks.

7.9.1.2 Data and Service Reliability and Availability

Reliability refers to the state when all the resources offered by the cloud service providers are up and running without any disruption. Moreover, reliability also ensures that the connection on which the different applications or services are operating is secure. However, the aspect of downtime poses critical problems for the reliability of cloud computing [33]. Redundant resource utilization is one of the ways in which reliability can be attained.

Despite all the prevailing facilities in the cloud architecture regarding high reliability and availability, the cloud model can experience various issues like denial-of-service attacks, slowdown of performance, breakdown of the equipment, and natural disasters. In a denial-of-service attack, the attacker shuts down a machine or network of a legitimate user by flooding it with traffic or by sending information to the system that can crash the system [34]. Besides, hardware availability is another matter of concern in cloud computing because if the required resources are not available, then it might give rise to disruption in the cloud network. Thus, it impedes all online activities of the customer.

7.9.1.3 Cryptography

The process of cryptography in cloud computing secures the information and data deployed on the cloud. This mechanism converts the data in plain text to ciphertext. It makes it impractical to comprehend the text unless a valid key is used to convert it back into a simple text. Therefore, the different cryptography methods should be implemented meticulously. The complete security of this method is reliant on an encryption key. While encrypting the data, key management is one of the most intricate security system issues and networks [3]. Key management involves protecting encryption keys from loss, unauthorized access, and corruption [35]. Many organizations do not implement encryption in their organizations because management is the most challenging task. Thus, the large number is broken down into smaller numbers, otherwise known as factorization, which bestows more security while encrypting the data using an RSA (Rivest Shamir Adleman) algorithm. Hence, the inappropriate implementation of RSA or using weak keys during encryption elevates the likelihood of the most prevalent brute force attack. Here, the adversary uses a trial-and-error method to guess the user's login details or the encryption keys or even find a hidden web page by attempting all possible combinations in the hope of making a correct guess. Even though this is an old type of attack, it is still effective and popular among hackers.

7.9.1.4 Cloud Data Recycling

Reusing the cloud space once the prevailing data was used adequately and sent to the garbage were considered a wise proposal. The process is referred to as sanitization, where a piece of data is cleaned and removed from a resource. The sanitization of data is quite a challenging task as it necessitates proper selection and disposition of the data that is to be sent to the garbage. The inadequate sanitization of the data might result in various severe issues such as leakage of data and data loss, as the hard disk of the cloud vendor might delete some significant data.

7.9.1.5 Data Backup and Recovery

Another noticeable issue in cloud computing is data backup and its recovery. To ensure the availability of data, there is a dire need to back up data on a regular basis that an organization might need to restore under unlikely circumstances to ascertain the smooth functioning of the business. The most common threat is if a nefarious third party gets unauthorized access, they will have the opportunity to access the entire information generated and owned by that particular organization [29]. The hackers can either tamper with the data or might even destroy the complete backup storage. The two vital aspects of cloud computing, i.e., resource pooling and elasticity, might create security issues. These features might jeopardize confidentiality and data recovery, as the resources that were allocated to the user who requested them might be given to another user. Consequently, the hacker would use the data recovery mechanism and then be able to find and access vital information from previous data, thus proving to be perilous for the sensitive data of the authorized user.

7.9.1.6 Security and Privacy

Data security is one such issue that is the consequence of the multi-tenancy model of cloud applications. This has always been a topic of discussion when adopting cloud services. As the services are shared in the cloud, the probability of a data breach increases to a great extent. There may be a situation where one user can easily get access to the data of another user, resulting in a serious problem. Undoubtedly, it is always daunting to place one's data and applications and execute the software of one's organization on the hard disk of a cloud provider company. In cloud computing, the data will be dispersed over different individual computers irrespective of the fact of where the base repository of the data is stored [13].

Privacy is one such area in cloud computing that is significantly affected. The user's data resides entirely on the premises of the cloud server, which elevates the privacy concern as the cloud vendors automatically get access to all the sensitive and confidential information of the organization [36]. They might inadvertently or purposely access the data available in the cloud and might tamper with the data either by changing it, removing, or misusing it. Consequently, efficacious identity theft leads to privacy loss and also greatly impacts the enterprise. The organization can experience short-term losses that may be due to remediation and investigation. In contrast, long-term issues might be due to the loss of integrity, confidence, and negative publicity.

7.9.2 Internet and Services Related Security Issues

The internet is considered to be one of the most vital components in cloud computing, just like the other cloud services and resources. The internet digitally transmits a large amount of packets from the sender to the receiver. This data is highly unsafe as it passes through a number of nodes in this process of transmission. There are many threats due to the exploitation of the internet.

7.9.2.1 Advanced Repeated Threats and Venomous Outsiders

In cloud computing, an advanced repeated threat (ART) is an attack that comprises of stages, namely, the information gathering stage, threats modeling phase, and the attacking stage. In the first stage, the hacker discovers all the crucial details regarding the server that is to be targeted. They then utilize public or intelligent private sources like open source intelligence (OSINT) to gather information about the target source. In the subsequent stage, the adversary attempts to locate the target server and also finds the best possible method of attack. Finally, the attacker in the last phase attacks the targeted server.

Hence, the first stage of this attack is the most significant phase for an attacker to succeed as it lets the attacker know on which server and how the attack has to be performed. Thus, making information public is a point of concern for business organizations [37].

7.9.2.2 Internet Protocols

Admittedly, the cloud infrastructure uses a variety of protocols for communication on the internet. As the cloud computing model has a web-based cloud infrastructure and uses numerous well-known vulnerable protocols, this might generate different network-based attacks. HTTP is one such delivery protocol that does not provide any assurance of the safe delivery of services. Besides, various web applications and HTTP protocols utilize a session handling technique that enables secure and unthreatened interactions between a single user and applications for the secure delivery of services. However, this technique is still not secure as it is vulnerable to session riding and session hijacking. Moreover, to create cookie poisoning and impersonation attacks, temporary cookies known as session cookies are used. Cookie poisoning, also known as session hijacking, refers to an attack strategy wherein the adversary either changes the data, produces a fraudulent copy of the same, hijacks it, or poisons a valid cookie that is sent back to a server. All these activities are done either to steal the data or bypass security.

7.9.2.3 Web Services

Data integrity is considered one of the major problems in a distributed system. Many SaaS vendors deliver services to their respective users via unreliable and unsafe application programming interfaces (APIs) that do not provide any transaction reports. Thus, the concern of data integrity surgers when several cloud services are offered to the consumers. Cloud services use web services description language (WSDL) to explicate the functionality given by web services. This re-engineering of data creates numerous issues like the man-in-the-middle attack or spoofing attack. In IP spoofing, the intruder alters the packet at the TCP level. This packet is utilized to attack the various systems connected to the internet that bestows different TCP/IP services [21]. Ideally, the attacker sends a packet that consists of an IP source address of a trusted and known host instead of its IP address to target the host to access the system of the legitimate user [31].

7.9.2.4 Web Technologies

A suitable web browser accesses almost all cloud services. There has been a consistent increase in malware in the cloud environment through the use of malicious internet links and websites. A considerable number of devices are connected to the web these days. This eventually proliferates the risk of cloud services getting attacked. A cross-site scripting (XSS) attack is considered one such common attack experienced in the cloud network. In this, malicious attacks are injected into certain websites and web-based applications to execute on the device of an unsuspecting victim. Besides, web technologies also face cookie theft and hidden field manipulation attacks wherein the intruder manipulates the concealed HTML fields as per his needs.

7.9.2.5 Service Availability

While accessing the cloud services on the web, the user comes across the most common denial-of-service (DoS) attack in which the perpetrator overloads the target machine with spurious service requests. This traffic congestion is performed to prevent reverting of the legal requests [38]. Cloud-based systems are vulnerable to DoS attacks as the cloud model offers its users one of its key features, resource pooling.

7.9.3 Network Security Issues

The fundamental component of the cloud computing model is the network. Cloud computing technology is dependent on the internet and remote computers to preserve the data for smoothly executing various applications. An efficient network is required to upload all the information to the cloud environment. In addition to this, cloud computing also offers its users facilities like virtual resources, high bandwidth, and different software as per their demand. Undeniably, the cloud network is highly vulnerable to numerous attacks and security threats detrimental to cloud technology. The nature of the cloud environment is quite dynamic so to disturb the availability of the services; an invader can perform a DoS attack. The DoS adversely affects the bandwidth of a particular network. It then proliferates the congestion in the targe network, resulting in the change in the network's boundary for the user's connectivity with the required service.

7.9.3.1 Mobile Networks

Employees of big companies sometimes utilize their electronic gadgets to access the applications of their organization. This refers to the concept of bring your own device (BYOD), which can at times be detrimental for enterprises. Even though this concept is advantageous for companies from the perspective of productivity, it brings forth many security threats when accessing cloud applications and services. The users use their smartphones to access the cloud-based SaaS applications and services that produce perilous malware. Threats like rooting and jailbreaking in mobile devices increase security threat as they access core parts of a device. The rooting facility in mobile phones facilitates the user to install fancy applications of their choice, and this facility also gives administration level permissions to the operating system [37]. This rooting process involves potential risks if it is done incorrectly, leading to security vulnerabilities as harmful applications get the opportunity to use the sensitive components of the operating system and therefore access the protected and confidential data. Another issue that arrives with using mobile platforms to access cloud services is data leakage. This not only leaks the personal data of the user but also discloses company data. Thus, many companies do not support the concept of BYOD because of the apparent issues that occur with the same.

7.9.3.2 Circumference Security

Circumference security is also referred to as perimeter security, wherein different techniques are employed at the network's perimeter to safeguard all the data and the resources. The foremost issue is the immobile network infrastructure; BYOD alters the prevailing security needs of the network. It also requires an open boundary-less network for different cloud services and applications. The biggest concern in the cloud environment is to attain security in the dynamic cloud network. Even though some standard and basic control mechanisms exist, they do not provide the complete security that is needed [37]. Besides, security threats while logging in and insufficient security concerns are other security problems with this approach.

7.9.3.3 XML Signature Element Wrapping

This is one of the most common and popular attacks witnessed on protocols. It uses the XML signature to verify and protect its integrity. This is a major attack on web services and also applies to the cloud computing environment. In this attack, the hacker attempts to modify or falsify a document signed digitally without tampering with the signature already in the node-set [39]. In other words, the adversary sends SOAP messages to the targeted host computer, which contains jumbled data that is not comprehensible by the user.

7.9.3.4 Browser Security

Entire computations are carried out on cloud servers in cloud models, whereas clients just send requests to the cloud server and then wait for the output. It is undeniable that a web browser is a common way to connect with cloud services. The web browsers utilize the Secure socket layer (SSL) process to encode the identity and credentials of the respective user [40]. Nonetheless, the SSL technology only assists point-to-point communications. This implies that if there is an intermediate between the cloud server and the client, like the firewall, then first the data must be decrypted on the intermediary host. This process eventually gives rise to a security issue on the web. The attacker might acquire all the user credentials from that intermediary host with the help of sniffing packages installed on the intermediate host. This signifies that the SSL technology is restricted in terms of its capabilities to provide appropriate authentication in cloud computing.

7.9.3.5 SQL (Structured Query Language) Injection Attack

Here the attacker inserts a malicious code into the SQL code to control the database server of the web applications in cloud computing. The hacker, through this attack, gets illegal access to the database along with the other confidential information [29].

7.9.4 Access Control Issues

Access control security is a technique that regulates the unauthorized read and write permissions of a resource in a cloud environment. This security measure minimizes the risk factor by authenticating data with a username and a password. The cloud model provides a multi-tenant feature to many customers wherein the cloud services and resources are accessible to the respective users either using a website or through a front-end interface like a web browser. Consequently, these two options to access the cloud services open the door to intruders and lead to different challenges.

7.9.4.1 Physical Access

A data center is a storehouse where data is stored at a centralized location. It provides numerous vital services to organizations, such as data storage, data backup and recovery, and management. Some organizations rent their data centers to users. There exist a plethora of access security issues in the cloud data centers, so no organizations can provide full physical security of the data. Data leakage in the data centers is considered to be the prime security concern. While developing the data center, the developer ensures all security measures to protect the information are taken. An organization's data is not only at risk from malicious outsiders, but an insider can also harm the data by accessing or controlling the internal physical security [36]. Sensitive information can be leaked either by a malicious insider or a malicious system admin who can access it and breach the user's privacy [34]. A cold boot attack is another way intruders steal crucial data by directly accessing the computer hardware. In other words, the hacker has physical access to the user's computer and then carries out a cold reboot to restart the machine with a motive to retrieve the encryption keys from the Windows Operating System. Besides, hardware tampering is yet another security issue in which the attacker intentionally implies an intrusion mechanism to perform an unauthorized physical or electronic action against a device.

7.9.4.2 User Credentials

Many multi-national companies utilize certain reliable technology like a lightweight access protocol or Microsoft Access Directory to manage users' credentials. Such protocols or directories are used for authentication and authorization. The access management server, which contains the credentials of all the personnel working in an organization, is kept either in the cloud location or within the organization itself using a firewall. As large companies consist of a large number of employees, so this eventually increases the load on the IT management team to add a new account, delete, or modify it. Also, it takes time to activate or deactivate an existing user account. Irrefutably, it is quite challenging to add a new credential to the access management directory. Weak credential recovery methods and credential reset may also lead to security issues in accessing the cloud services. The cloud service providers are always worried about the user login credential security. If they are stolen, then sensitive and crucial data can be accessed by the attacker. They can also hamper the data by easily monitoring and manipulating it. If the credentials are stolen, then the user might encounter various issues like phishing attacks, man-in-the-middle attacks, replay attacks, and session hijacking. The most commonly witnessed attack in the SaaS model of cloud computing is the man-in-the-middle attack. The hacker obstructs and alters the communication between the legal user and the server without bringing it to the client's knowledge [21]. On the other hand, in the session hijacking, the session of the web user is taken over secretively by an attacker. The hacker gets the user's session ID by unfair means and then impersonates it as an authorized user [30]. The intruder invades the system surreptitiously and accesses the network just like a legitimate user.

7.9.4.3 Entity Authentication

To securely access or use the cloud-based applications, various robust authentication techniques are required. However, weak authentication approaches can result in numerous cloud attacks. Like, in the brute force attack, the attacker tries all the possible permutations and combinations with letters, numbers, and alphanumeric characters until the correct password gets discovered [41]; while in the dictionary attack, the intruder tries to gain illegitimate access to the system by entering common words and phrases of a dictionary [38].

7.9.5 Software Security Issues

In modern times, people write software programs by using different programming languages that consist of thousands or even millions of lines of code. Because of the complexity that prevails in programming, software security issues can be encountered in a system.

7.9.5.1 Platforms and Frameworks

Admittedly, the cloud model provides users with the feature of resource pooling. The PaaS model offers its users a shared platform to deploy cloud applications. Moreover, it also supports various languages that help in developing various cloud applications. The cloud platforms are also not devoid of issues.

The method of creating isolation also has certain limitations. The most obvious method is by using a discrete java virtual machine (JVM) for each program, but this method also has its limitations. It consumes memory in bulk and is also not secure. On the contrary, Java capabilities are another way to create isolation. Even though this method isolates one class from another, this approach is also not devoid of security threats. It does not prevent sensitive data and information leakage and safe thread termination.

7.9.5.2 User Front-End

Using a standard web browser on the internet, the user accesses various infrastructure (IaaS) and software (SaaS) services bestowed by the cloud vendors. While accessing the cloud services, an interface acts as a gateway via the internet. The attacker thus uses this interface to intrude into the system. The hacker disrupts the security firewall and barriers of the system by various inappropriate means, including imperfect configurations, unauthorized access, or injecting masked code into the system. In the cloud injection attack, the intruder's goal is to remain camouflaged from the user by inserting a malicious service or virtual machine into the system, which looks like a legitimate service to the user, just like the one already running on the cloud [42]. If the attacker succeeds, then the cloud system will treat the malicious content as valid; consequently, the server redirects all the valid user requests to the malicious server.

REFERENCES

1. L.Wang, G.von Laszewski, A.Younge, X.He, M.Kunze, J.Tao, and C.Fu, "Cloud Computing: A Perspective Study," *New Generation Computing*, vol. 28, no. 2, pp. 137–146, 2010.
2. P.Mell, and T.Grance, "The NIST Definition of Cloud Computing," *National Institute of Standards and Technology (NIST)*, vol. 800, no. 145, 2011.

3. M.Carroll, A.van der Merwe, and P.Kotze, "Secure Cloud Computing: Benefits, Risks and Controls," *2011 Information Security for South Africa*, pp. 1–9, 2011.
4. M. N. O.Sadiku, S. M.Musa, and O. D.Momoh, "Cloud Computing: Opportunities and Challenges," *IEEE Potentials*, vol. 33, no. 1, pp. 34–36, 2014.
5. R.Piplode, and U. K.Singh, "An Overview and Study of Security Issues & Challenges in Cloud Computing," *International Journal of Advanced Research in Computer Science and Software Engineering*, vol. 2, no. 9, pp. 115–120, 2012.
6. S.Zhang, S.Zhang, X.Chen, and X.Huo, "Cloud Computing Research and Development Trend," *2010 Second International Conference on Future Networks*, pp. 93–97, 2010.
7. M.Malathi, "Cloud Computing Concepts," *2011 3rd International Conference on Electronics Computer Technology*, vol. 6, pp. 236–239, 2011.
8. T.Dillon, C.Wu, and E.Chang, "Cloud Computing: Issues and Challenges," *2010 24th IEEE International Conference on Advanced Information Networking and Applications, Pr*, pp. 27–33, 2010.
9. T.Diaby, and B. B.Rad, "Cloud Computing: A Review of the Concepts and Deployment Models," *International Journal of Information Technology and Computer Science*, vol. 9, no. 6, pp. 50–58, 2017.
10. J.Srinivas, K. V. S.Reddy, and A. M.Qyser, "Cloud Computing Basics," *International Journal of Advanced Research in Computer and Communication Engineering*, vol. 1, no. 5, 2012.
11. A.Rashid, and A.Chaturvedi, "Cloud Computing Characteristics and Services a Brief Review," *International Journal of Computer Sciences and Engineering*, vol. 7, no. 2, pp. 421–426, 2019.
12. I.Odun-Ayo, M.Ananya, F.Agono, and R.Goddy-Worlu, "Cloud Computing Architecture: A Critical Analysis," *2018 18th International Conference on Computational Science and Applications (ICCSA)*, pp. 1–7, 2018.
13. M. B.Mollah, K. R.Islam, and S. S.Islam, "Next Generation of Computing through Cloud Computing Technology," *2012 25th IEEE Canadian Conference on Electrical and Computer Engineering (CCECE)*, pp. 1–6, 2012.
14. Y.Jadeja, and K.Modi, "Cloud Computing – Concepts, Architecture and Challenges," *2012 International Conference on Computing, Electronics and Electrical Technologies (ICCEET)*, pp. 877–880, 2012.
15. L.Wang, J.Tao, M.Kunze, A. C.Castellanos, D.Kramer, and W.Karl, "Scientific Cloud Computing: Early Definition and Experience," *2008 10th IEEE International Conference on High Performance Computing and Communications*, pp. 825–830, 2008.
16. S. V.Nandgaonkar, and A. B.Raut, "A Comprehensive Study on Cloud Computing," *International Journal of Computer Science and Mobile Computing*, vol. 3, no. 4, pp. 733–738, 2014.
17. S.Alshomrani, and S.Qamar, "Cloud Based E-Government: Benefits and Challenges," *International Journal of Multidisciplinary Sciences and Engineering*, vol. 4, no. 6, pp. 1–7, 2013.
18. S. P.Mirashe, and N. V.Kalyankar, "Cloud Computing," *Journal of Computing*, vol. 2, no. 3, pp. 78–82, 2010.
19. S.Patidar, D.Rane, and P.Jain, "A Survey Paper on Cloud Computing," *2012 Second International Conference on Advanced Computing & Communication Technologies*, pp. 394–398, 2012.
20. K.Stanoevska-Slabeva, and T.Wozniak, "Cloud Basics – An Introduction to Cloud Computing," in *Grid and Cloud Computing: A Business Perspective in Technology and Applications*, Heidelberg, Berlin: Springer, 2010, pp. 47–61.

21. L.Savu, "Cloud Computing: Deployment Models, Delivery Models, Risks and Research Challenges," *2011 International Conference on Computer and Management (CAMAN)*, pp. 1–4, 2011.
22. A.Gajbhiye, and K. M.Shrivastva, "Cloud Computing: Need, Enabling Technology, Architecture, Advantages and Challenges," *2014 5th International Conference – Confluence the Next Generation Information Technology Summit (Confluence)*, pp. 1–7, 2014.
23. Z.Mahmood, "Cloud Computing: Characteristics and Deployment Approaches," *2011 IEEE 11th International Conference on Computer and Information Technology*, pp. 121–126, 2011.
24. F. F.Moghaddam, M. B.Rohani, M.Ahmadi, T.Khodadadi, and K.Madadipouya, "Cloud Computing: Vision, Architecture and Characteristics," *2015 IEEE 6th Control and System Graduate Research Colloquium (ICSGRC)*, pp. 1–6, 2015.
25. M. G.Avram, "Advantages and Challenges of Adopting Cloud Computing from an Enterprise Perspective," *Procedia Technology*, vol. 12, pp. 529–534, 2014.
26. R.Bhadani, "A New Dimension in HRM: Cloud Computing," *International Journal of Business and Management Invention*, vol. 3, no. 7, pp. 13–15, 2014.
27. A.Albugmi, M. O.Alassafi, R.Walters, and G.Wills, "Data Security in Cloud Computing," *2016 Fifth International Conference on Future Communication Technologies (FGCT)*, pp. 55–59, 2016.
28. M.Ahmed, and M.Ashraf Hossain, "Cloud Computing and Security Issues in the Cloud," *International Journal of Network Security & Its Applications*, vol. 6, no. 1, pp. 25–36, 2014.
29. R.Jathanna, and D.Jagli, "Cloud Computing and Security Issues," *International Journal of Engineering Research and Applications*, vol. 7, no. 6, pp. 31–38, 2017.
30. P.Mosca, Y.Zhang, Z.Xiao, and Y.Wang, "Cloud Security: Services, Risks, and a Case Study on Amazon Cloud Services," *International Journal of Communications, Network and System Sciences*, vol. 7, no. 12, pp. 529–535, 2014.
31. S. M.Shariati, Abouzarjomehri, and M. H.Ahmadzadegan, "Challenges and Security Issues in Cloud Computing from two Perspectives: Data Security and Privacy Protection," *2015 2nd International Conference on Knowledge-Based Engineering and Innovation (KBEI)*, pp. 1078–1082, 2015.
32. L.Kacha, and A.Zitouni, "An Overview on Data Security in Cloud Computing," *Advances in Intelligent Systems and Computing*, vol. 661, pp. 250–261, 2017.
33. M.Sajid, and Z.Raza, "Cloud Computing: Issues & Challenges," *International Conference on Cloud, Big Data and Trust*, vol. 20, no. 13, pp. 13–15, 2013.
34. Y. Z.An, Z. F.Zaaba, and N. F.Samsudin, "Reviews on Security Issues and Challenges in Cloud Computing," *IOP Conference Series: Materials Science and Engineering*, vol. 160, pp. 1–9, 2016.
35. A.Mondal, S.Paul, R. T.Goswami, and S.Nath, "Cloud Computing Security Issues & Challenges: A Review," *2020 International Conference on Computer Communication and Informatics (ICCCI)*, pp. 1–5, 2020.
36. B. S.Al-Attab, and H. S.Fadewar, "Security Issues and Challenges in Cloud Computing," *International Journal of Emerging Science and Engineering*, vol. 2, no. 7, pp. 22–26, 2014.
37. A.Singh, and K.Chatterjee, "Cloud Security Issues and Challenges: A Survey," *Journal of Network and Computer Applications*, vol. 79, pp. 88–115, 2017.
38. S.Naik, "Authentication Attacks in Cloud Computing: A Survey," *International Journal of Technology Research and Management*, vol. 2, no. 5, pp. 1–7, 2015.
39. M.Jensen, J.Schwenk, N.Gruschka, and L. L.Iacono, "On Technical Security Issues in Cloud Computing," *2009 IEEE International Conference on Cloud Computing*, pp. 109–116, 2009.

40. D.Jamil, and H.Zaki, "Security Issues in Cloud Computing and Countermeasures," *International Journal of Engineering Science and Technology*, vol. 3, no. 4, pp. 2672–2676, 2011.
41. S. K.Sood, "A Combined Approach to Ensure Data Security in Cloud Computing," *Journal of Network and Computer Applications*, vol. 35, no. 6, pp. 1831–1838, 2012.
42. A.Singh, and M.Shrivastava, "Overview of Security Issues in Cloud Computing," *International Journal of Advanced Computer Research*, vol. 2, no. 1, pp. 41–45, 2012.

8 Intrusion Detection Systems for Trending Cyberattacks

Shahbaz Ahmad Khanday, Hoor Fatima, and Nitin Rakesh

CONTENTS

8.1 INTRODUCTION

Intrusion detection systems are used to test and examine inward and outward moving traffic for distinguishing malicious network traffic from the normal network traffic flow (Figure 8.1). After detecting problematic network traffic patterns, alarm signals are sent to the console or administrator, when such disturbances are monitored on a network. The information and events are recorded using a detection system to report on any malicious activity or violation. Some intrusion detection systems can respond to intrusions as soon as they are discovered. Any harmful invasion or breach

DOI: 10.1201/9781003229704-8

is forwarded to a console or an administration managing the violation requests using a centralized SIEM (security information and management system) system. A SIEM system uses alert filtering algorithms to discriminate between malicious and false alerts by combining data from several sources. Although intrusion detection systems monitor networks for suspicious activity, they are prone to false alarms. As a result, when enterprises first implement IDS products, they must fine-tune them. This entails properly configuring intrusion detection systems to distinguish between legitimate network traffic and malicious activities. Intrusion prevention systems perform similar tasks to intrusion detection by examining the packets and frames approaching and entering the network area and to look for harmful or infected activity in the network and devices, and provide warning signals immediately.

8.2 CLASSIFICATION OF INTRUSION DETECTION SYSTEMS

Types of intrusion detection system are distinguished on the basis of characteristics utilized to detect intrusions and misbehaviors in various segments of a network. All the kinds of IDS are determined by the topological placement and presence on sections of the network. Figure 8.2 sets out the details.

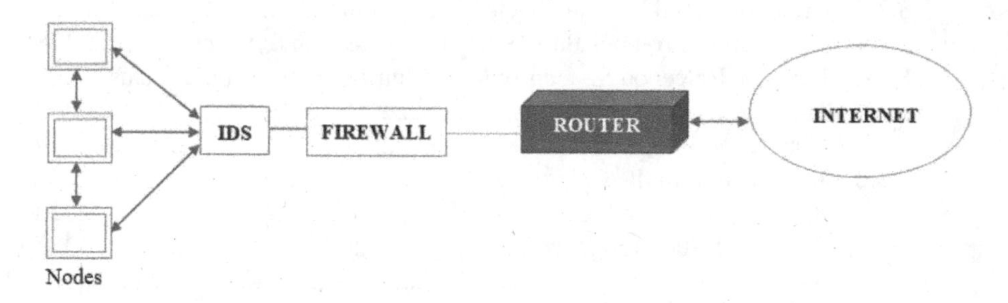

FIGURE 8.1 Intrusion detection system [1].

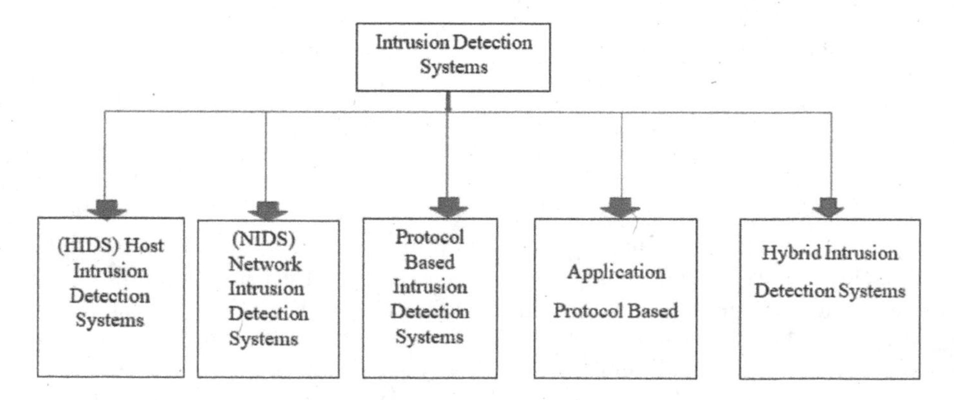

FIGURE 8.2 Classification of intrusion detection systems.

8.2.1 Host Intrusion Detection Systems

HIDSs examine information found on a single computer or several host computers, including operating system contents and application files (Figure 8.3). NIDS (network intrusion detection systems) need to be installed at a predetermined location within the vicinity of the network to classify the traffic from devices that are part of the network [1]. They monitor all subnet traffic and compare it to a list of known threats. The alert can be sent to administrators once an assault is detected or strange activity is found [2].

8.2.2 Network-Based Intrusion Detection Systems

Data gathered via network communications is evaluated by NIDSs (Figure 8.4). Network intrusion detection systems that work on various hosts or devices are known as HIDSs (host intrusion detection systems). Only the device's incoming and outgoing packets are monitored by a HIDS, which notifies the administrator if anything problematic or harmful behaviors to the system are found. It uses the existing state of the network or system as a reference and compares it with the current state. A warning is triggered if the system files important for analyzing are modified or destroyed. In most network architectures and topologies NIDS are located on the server side and away from the individual nodes of the network, where the data traffic flowing throughout the network can be monitored [3].

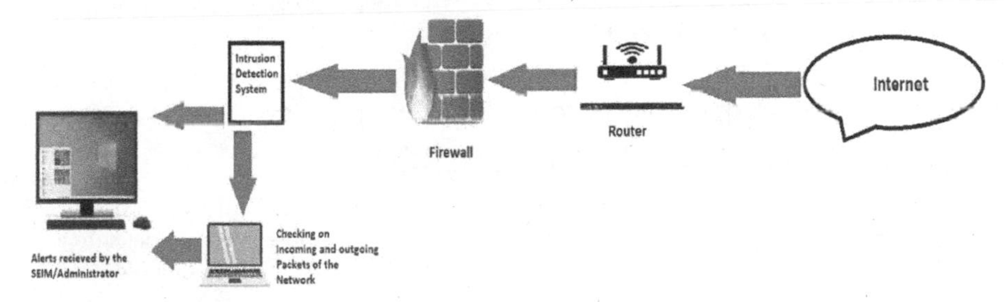

FIGURE 8.3 Host intrusion detection systems.

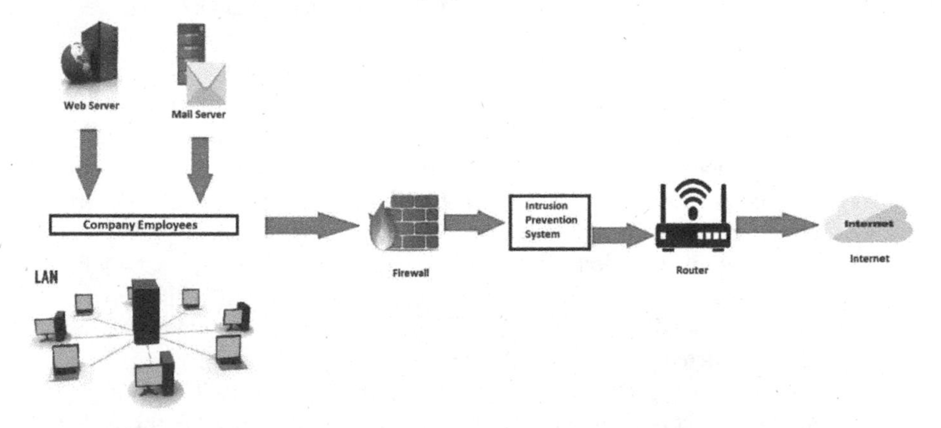

FIGURE 8.4 Network intrusion detection systems.

8.2.3 Protocol-Based Intrusion Detection Systems

This kind of intrusion detection system sits at the front end of a server as an monitoring and communication tool employed by the server and client protocols suites. PIDSs monitor and check the protocol's performance and circumstances [3]. They make an attempt to protect the web server by general inspection of the data streams of the HTTPS protocol and by accepting all the HTTP protocol requests linked with it. Because HTTPS isn't encrypted and isn't available right away, it's a bad idea to use it.

8.2.4 Application Protocol-Based Intrusion Detection Systems

An application protocol-based intrusion detection system (APIDS) is a type of intrusion detection mechanism or approach that typically resides between multiple servers (APIDS). It analyzes and assesses application-specific communication protocols to detect intrusions. This would, for example, test the performance of the SQL protocol as the intermediate transacts. Application protocol-based intrusion detection systems are designed for when multiple servers are grouped for traffic monitoring against different kinds of intrusions. ABIDS (program-based intrusion detection system) is one kind of HIDS that evaluates events inside a software application. The application's transaction log file is the most typical information source for an Application-Based IDS.[7]

8.3 HYBRID INTRUSION DETECTION SYSTEMS OR MIXED INTRUSION DETECTION SYSTEMS

Hybrid or mixed intrusion detection systems are a kind of IDS that uses a host IDS and a network IDS, or any combination or mixture of two or more intrusion detection systems, to obtain and complete an accurate detection [3]. However, evaluating data with a MIDS takes a long time.

Intrusion detection systems can be further classified on the basis of the detection mechanism. The different kinds of methods used for the detection of intrusions are:

- Signature-based intrusion detection – A signature-based IDS identifies attacks using specified patterns in network traffic, such as the number of bytes or the number of 1s or 0s. It also recognizes malware based on previously identified malicious instruction sequences. Signatures are the recognized patterns in the IDS. Signature-based IDS can quickly identify attacks based on their signature (pattern).
- Anomaly-based intrusion detection systems – As we are witnessing the rapid and immense growth in malware attacks on different network architectures and devices, there is a requirement for an intrusion detection system that detects malware [9]. Anomaly-based IDSs are designed to identify unknown malware threats. Machine learning is used in anomaly-based IDSs to construct a trustworthy activity model, and anything that comes in is compared to that model and considered suspicious if it is not

found in the model. In comparison to signature-based methods, the machine learning-based method offers greater generalized protection [16].

- Specification-based intrusion detection – A SIDS is also known as a state-full protocol analysis. Specification-based intrusion detection systems or state-full protocol analyses are protocol-oriented, where the state of the traffic is monitored and inspected at different stages on the network and transport layer. After analyzing and examining the protocols suites, the variation in commands or command sets can be a key characteristic for comparing with normal behavior. State-full protocol inspection is comparable to anomaly-based detection in that it can examine traffic on the network and transport layers, as well as vendor-specific traffic on the application layer, which AIDS, in comparison with specification-based protocols, cannot.

8.4 FIREWALLS

A firewall on a network is an essential component to ensure security; it monitors all incoming and outgoing traffic and accepts, rejects, or drops that traffic on the basis of some predefined security principles (Figure 8.5). Firewalls can be used as an important asset to prevent intrusions, where they can provide flexibility and guidance to the incoming packet stream or completely ignore the stream [10, 11, 12, 13, 14]. A firewall is a device, either software or hardware, that creates a barrier between the internal network architecture and external sources or networks, which cannot be secured by any means, such as the internet [8]. Traffic from outside the network, like the internet, is a vital source for any kind of misbehavior or intrusion. Some operations performed by a firewall are:

- Allow – Indicates a green signal and permits the incoming traffic.
- Reject – When a firewall detects anything problematic to a network, it does not allow the traffic to pass and reach the nodes and users of the network. In this case, the firewall sends a rejection acknowledgment/message to the sender.
- Drop – The drop operation by a firewall is quite similar to the reject operation, but in drop operation the incoming traffic is not allowed to go through and does not send an acknowledgment or error message to the source.

8.4.1 Classification of Firewalls

Firewalls can be classified into multiple different groups, where the behaviors of operation and generation can validate the type of firewall. Firewalls can be categorized into certain groups (Figure 8.6):

- *Packet filter firewalls*

  Packet filter firewalls examine the traffic of packets moving in and out of a network on the basis of the attributes of the packet [8, 9]. Each packet

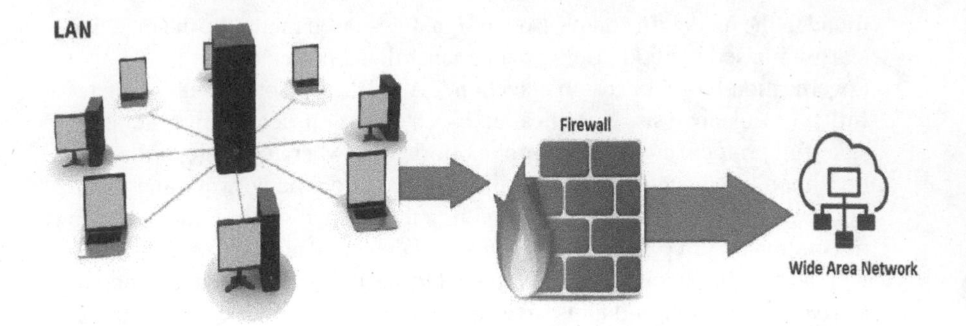

FIGURE 8.5 Firewalls.

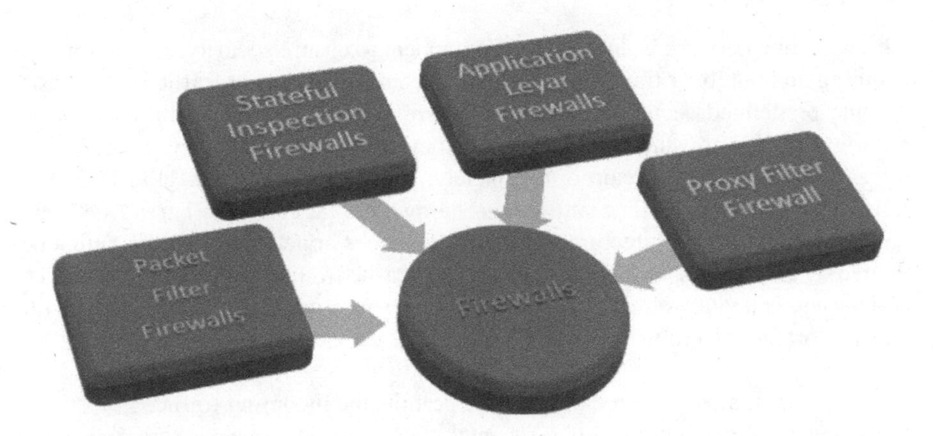

FIGURE 8.6 Classification of firewalls.

is considered a valuable entity and the properties of every packet are used to determine whether the packet should be allowed to pass or dropped.

Properties of a packet, like a source IP, destination IP, source port number, destination port number and protocol suite, are important attributes for a packet filter firewall to decide what to do with the packet. The packet filtering firewall keeps track of a filtering table that determines whether a packet is forwarded or deleted [8]. The packets will be filtered using the following rules when a packet filtering table is maintained by a firewall. The table and the rule set given in Table 8.1 is an example of a filtering table.

Table 8.1 depicts the rule set for a packet filter firewall when there are packets flowing toward the network secured by a packet filter firewall. The rule set in the table assists the firewall to allow trustworthy packets in and deny packets from sources which are labeled as "deny." In the meantime, the firewall works according to the rule set defined in the table.

TABLE 8.1
Packet Filtering Table

S.NO	Source IP address	Destination IP address	Source port number	Destination port number	Action
1.	122.2.12.1	-----	------	-------	deny
2.	-----	------	--------	23	deny
3.	-----	122.21.11.4	-----		deny
4.	-----	122.21.11.1	-----	1021	allow

- *Stateful Inspection Firewall*

  Unlike packet filtering firewalls, stateful firewalls execute stateful packet inspections, which can determine packet connection state, making them more efficient. It monitors the status of network connections, such as TCP streams, that pass through it. As a result, filtering decisions are based not just on set rules but also on the packet's state table history.
- *Application Layer Firewall*

  Any OSI layer, up to the application layer, can be inspected and filtered by an application layer firewall. The firewall may restrict particular material and detect when certain applications and protocols (such as HTTP and FTP) are being abused. One example is a proxy filter firewall.

  Any effort to obtain unauthorized access to a computer, computing system, or computer network with the purpose of destruction is considered a cyberattack. Cyber assaults attempt to damage, interrupt, ruin, or seize command of computer systems, as well as to change, block, erase, modify, or steal the data stored on them. Any individual or organization may conduct a cyber-assault from anywhere using one or more attack techniques. Cybercriminals are those who carry out cyber assaults. Individuals who act alone, depending on their ability to develop and execute harmful assaults are known as bad actors, threat actors, and hackers. They might also be members of a criminal gang that collaborates with other threat actors to identify flaws or issues in computer systems, known as vulnerabilities, that can be leveraged for financial benefit [17].

8.4.2 Introduction to the Cyberattacks

Technology has taken over the world. Several emerging innovations have been invented since the Industrial Revolution that have aided in the improvement of living. Since the 1980s, the usage of computers has been the most significant technological advancement. Computers have evolved from huge, complicated equipment to user-friendly, interactive gadgets that everyone can use [18]. Computers, when combined with the internet, have made communication much simpler. Computers and the internet play an important role in modern civilization. The usage of the internet has resulted in the creation of a virtual communication space known as cyberspace, in which information is sent by one kind of guided media type (fiber optic cables) or other types of wires using different internet technologies. As more information is put into it, the size of this area has gradually grown. Banking, hospitals, education, emergency services, and the military have all become more reliant on cyberspace. The level of difficulty for protecting against attacks has risen as well. Cyberattack is the term for such dangers. The scope, depth and intensity of these attacks have been rising over a number of years. Currently there is a considerable dearth of information regarding the many types of assault, their frequency of propagation, and their respective intensity [19].

As a result, many organizations and governments have become exposed to such attacks. Successfully implementing security measures necessitates a detailed grasp of these types of threats and how they are classified. As a result, a thorough list of cyber assaults and their categories is an important part of cyber-security efforts. The research tries to categorize attacks based on various character traits such as intensity, intention, and lawfulness in order to provide software developers with a greater grasp of the motivation behind such assaults, which may contribute to creating security systems and processes depending on the type of attack.

A computer network assault, often known as a cyberattack, is defined as a disturbance of the security or validity of data or information. Hacking entails scouring the internet for systems with weak security controls and systems that are compromised, and then inserting malicious code that produces mistakes in the output by changing the logic of the program. Once a hacker has infiltrated a system, he or she may remotely access it and send orders to make it function as a spy for the attackers as well as damage other systems. The attacker will exploit vulnerabilities in the infected system, such as software defects, anti-virus deficiencies, and faulty system settings, so that additional systems may be compromised through it. The goal of a cyber-assault is to breach the security policies of a system or to get control, and retrieve information/data, sometimes gaining privileges to modify the data of a company or government entity. To acquire data or information, the attacker or hacker must exhibit specific qualities in order to succeed [17]. The characteristics can be seen in Figure 8.7.

8.4.2.1 Integrated

In order to infect the system, the attacker will require the process to be synchronized. They achieve what they want by synchronizing the procedures required to steal the information. The hackers will receive their results on time, in step, and in line with their expectations.

8.4.2.2 Structured

The attacker or hacker will utilize a well-organized set of behaviors to infect the system with ease. They get more efficient results by employing rationally ordered approaches.

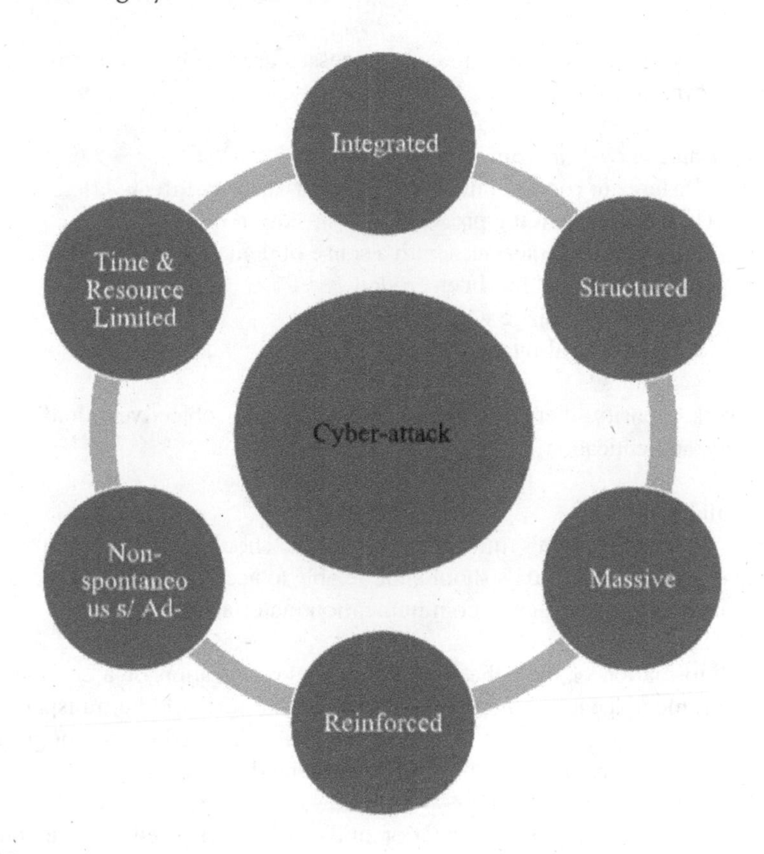

FIGURE 8.7 Cyberattack characteristics.

8.4.2.3 Massive

When assaults are launched, they are generally large-scale, infecting billions of machines throughout the world and inflicting massive data and financial damage.

8.4.2.4 Reinforced

The attacks are meticulously planned and executed in such a way that the ensuing damage is substantial enough to jeopardize the organization's operations.

8.4.2.5 Not Spontaneous or Ad Hoc Attacks

These are planned meticulously and systematically in order to wreak maximum damage.

8.4.2.6 Time and Resources are Limited

Attacks will be planned from the beginning, so it takes a long time and a lot of money to put one together.

Official sites, financial organizations' websites, online discussion forums, news and media websites, and military/defense networks' websites are the most common targets of cyber assaults. The following processes are involved in the purpose and motives of cyberattacks [17]:

- *Information Obstruction*
  - Measures to counter international cyber-security threats.
  - The decision-making process is being slowed down.
  - Providing public services with a sense of denial.
  - The public's trust has been eroded.
  - The country's image will be tarnished.
  - Displacing legal interests.

For network security, there are five primary security objectives: confidentiality, availability, authentication, integrity, and non-repudiation.

- Confidentiality

  Any organization's information or data should be kept secure, and unauthorized individuals should not be able to access it readily. In terms of security, secure storage of communication material is critical.
- Availability

  Information or data that is critical to the operation of a company or government agency should be kept hidden, but it should be transparent to authorized users and difficult to access by unauthorized users. For genuine users, various restrictions must be implemented.
- Authentication

  Before accessing the information or data, the authorized users' identities should be confirmed. There are three options for confirming a valid user's identity: passwords, tokens, and biometrics. It is simple to distinguish between authorized and unauthorized users using these verification methods.
- Integrity

  During communication, the information or data should not be changed. The data must arrive at the destination exactly as it was transmitted from the origin.
- Non-repudiation

  The parties transmitting and receiving the data or information should guarantee that they are the entities involved in the communication, and must have knowledge of the lag in transmitting and receiving the data frames or information [20].

  The intensity of the cyber assaults, as well as their involvement, are classified further (Figure 8.8). They are [21, 22]:

1. Active Attacks – In active attacks, the attacker has the ability to broadcast data frames to all entities within his vicinity, or the victim may be a single target. Because the attacker is positioned between the intercommunicating parties, he or she may attempt to exclude the target from services like sending and reception of data inside the network.

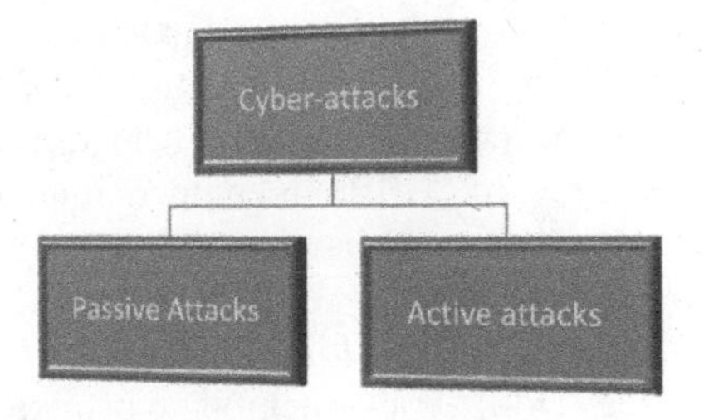

FIGURE 8.8 Classification of cyberattacks.

Because the server cannot validate the relevant data without validating the knowledge obtained, the attacker next tries to spoof the client after the authentication procedure is finished. In active attacks, the attacker can modify intercepted data streams.

2. Passive Attacks – An unauthorized attacker wiretaps or intercepts communication between two parties in order to steal information stored in a system. This is distinct from an active assault in that it doesn't include tampering with the database, though it could still be criminal.

8.5 TRENDING CYBERATTACKS

Since the majority of the world's population has shifted to internet-based labor, India, one of the world's largest countries with a population of around 136.64 million people, has followed suit (1.36 billion) [4–6]. During the pandemic, there was a significant increase in cyberattacks in India. Attackers attempted to obtain access to privileges and information from individuals who were in pandemic lockdown conditions.

- ***Malware***

  Malware is a broad word that encompasses a wide range of harmful software types, including software flaws, ransomware, and spyware. Malware refers to harmful software that is often composed of code developed by cybercriminals. The creation of malware is a serious threat to worldwide information security. According to recent research by Microsoft [15], every third malware scan resulted in positive identification. The IoT networks have also fallen victim to the modern era of cyberattacks, with most network technology development being a result of new IoT paradigms [28, 29]. Some common examples of IoT networks under threat are smart homes, health care services and equipment, smart grids, and much more [30].

  Anti-malware software that is up to date and consistent with current malware development is therefore critical to the survival of computers nowadays. This has proven to be a challenging project. Malware has evolved into a viable technique for doing illegal business, and its creators go to considerable efforts to prevent detection by anti-malware software.

As a result, new malware variants appear at an alarmingly quick rate, with some malware families comprising tens of thousands of variants [16]. Trend Micro Research has discovered coronavirus-themed malware that disables a system's master boot record (MBR) and renders it unbootable.

The virus was described and discussed (NUKIB) [17] in a research report published by the Czech cyber-security agency.

- ***Denial-of-Service/Distributed Denial-of-Service***

  A denial-of-service (DoS) attack attempts to bring a system or network to a stop, leaving it inaccessible to the intended audience. DoS attacks operate by flooding the target with requests or providing it with information that causes it to fail. In all situations, the DoS attack depletes the facility or asset intended by legitimate customers (i.e., employees, owners, or account holders).

  One of the most well-known and dangerous cyberattacks nowadays is the distributed denial-of-service (DDoS) attack. Its modest yet extremely efficient attack techniques pose a serious threat to the contemporary internet ecosystem [18]. To prevent DDoS attacks, several pieces of research have been done. However, in today's environment, putting in place or developing relevant systems like intrusion protection, intrusion detection systems, and other procedures, DDoS attacks are constantly emerging and upgrading. These types of investigations are vulnerable to DDoS assaults.

  Rather, its scope and intensity are expanding. As a result, it's critical to figure out why these DDoS defense studies aren't halting and deterring attacks. The increase is ascribed to an increase in web traffic as a result of the COVID-19 pandemic, with many businesses relying on internet technologies and e-commerce and other relevant approaches to stay alive [16].

- ***Hacking***

  Hacking is the act of attempting to infiltrate electronic devices such as laptops, cell phones, tablets, computers, and even whole networks. Hacking is a term used to describe an attempt to manipulate a computer or a hidden network within a computer. Simply said, it is the unauthorized access or command of cyber-security systems with the purpose of committing a crime. Hacking is an offense that consists of breaking into a system or a network to attain almost complete control over it.

  Nowadays, hackers are targeting national health care and science centers in the United States, France, Spain, Thailand, and other countries, prompting the latest warning of cyber assaults during the pandemic. The majority of locations were affected with a kind of ransomware, a type of malware that blocks the users from accessing their own resources, which meanwhile turns out to be encrypted data. Others had network outages or data breaches and other network issues [20].

- ***Phishing***

  Phishing is a sort of cyberattack in which a person pretending to be a real person contacts a victim or targets them by email, phone, or text message in an attempt to get sensitive information, such as personal information, banking and credit card details, and credentials. The matter of fact involves the fooling of a smooth prey, which can act and interpret according to the system or person carrying out the phishing attack. This places a variety of organizational societies under threat.

Hackers employ an array of different techniques to create disorder on the internet and deceive a huge number of people in a variety of ways. As a result, it's critical for people in these groups to know how to protect themselves when online or to deal with cyber-related technologies [21]. Phishers have sent emails purporting to offer expert advice on how to protect against coronavirus.

- ***Ransomware***

  The ransomware virus has been spreading like a cyclone. A cyclone wind creates turbulence in the environment, just like ransomware does for computer data. Any consumer that is gravitating toward digital technologies needs to keep his or her data protected. So, what happens if information is stolen? Ransomware is a type of computer virus that encrypts its victims' data.

  An effective assault has far-reaching consequences that go far beyond the expense of the ransom. Customers may be inconvenienced, and data may be permanently destroyed, resulting in lost efficiency, lost business, and consumer dissatisfaction [22]. Remote working has already been shown to greatly raise the probability of an effective ransomware attack. This rise is associated with a number of home IT monitors and a greater chance of consumers clicking on COVID-19-themed ransomware emails due to their anxiety levels.

- ***Financial frauds***

  Financial manipulation is becoming more prevalent, posing a significant danger to the financial industry. As a consequence, financial firms are compelled to develop their risk mitigation processes on a regular basis. A few researchers have used data mining and machine learning methods to find alternatives to this challenge over the past months [23, 24].

  The COVID-19 virus is now wreaking havoc on people all over the world. The number of infected individuals grows by the hour, and mobility controls are tightening in many nations, health services are failing, and financial markets have lost the most money since the economic collapse of 2008 [25]. The pandemic has seen financial services becoming one of the most hit industry by cyberattacks.

- ***Email Spamming***

  Over the last two years, the epidemic of spam emails has increased. It is not only inconvenient for consumers, but also harmful to those who fall victim to fraud and other assaults. This is owing to the sophistication of email spamming tactics, which are progressing from conventional spamming (direct spamming) to a more modular, ambiguous, and indirect method using botnets for distributing email messages [26].

- ***Extortion***

  Though email viruses and worms are well-known, they aren't the only cyber threat that companies should be worried about when it comes to management and data security. New web extortion techniques are becoming more common, prompting investigators to ask if organized crime is to blame in some instances. When the world responded to COVID-19, cybercriminals saw a chance to use the pandemic to spread malware. People all over the world were expected to work remotely, which shifted the cyber threat environment significantly as malicious hackers saw an opening to exploit remote technology [27].
- ***Crypto-jacking***

  Crypto-jacking is a type of cyberattack that infiltrates a computer or mobile device and then mines any information related to the cryptocurrency using the device's resources. Malicious adversaries use web users' CPU resources to mine cryptocurrency information by injecting malicious payloads onto infiltrated websites, as per the new research [31]. Crypto-jacking can be well understood as crypto-mining programs that include worming abilities, which allows them to infect other network devices and servers. This makes them more difficult to spot and eradicate. These scripts may indeed test to see if the device has previously been attacked with crypto-mining malware from a competitor. The script disables another crypto-miner if one is identified [32].
- ***Social media scams and privacy issues***

  Social media scams are also popular methods used by attackers to legitimate social network users. In these kinds of scams, social media is used to send a specially constructed message or post, which somehow gains the attention of a social network user, and he/she ends up being a part of the event. Some examples of social networking scams are advertisements, endorsements, investment advertising, and sometimes charity functions or events, all asking for donations. Individuals affected by illegal behavior may be unaware of the dangers of leaking data and knowledge, failing to find out how to deal with the issues and handle them in certain situations [33]. Accepting and intervening in the scams can be lethal to a user; in many the user can lose credentials and could be a victim of financial fraud.

  The privacy issue when using social media has been one of the prime issues on modern social networking websites like Facebook, Twitter, and Instagram, etc.; the user is afraid of privacy and confidentiality issues, when it comes to sharing data on public platforms [34].
- ***Social engineering***

  Social engineering is the practice of persuading someone to provide sensitive or personal information that may be exploited for fraudulent reasons, generally through digital communication. Physical, sociological, and technological components of social engineering assaults are employed at various phases throughout the attack [35]. When an attacker wishes to join to an organization's network, this is a simple example of a social engineering assault. As a consequence of his investigations, the attacker discovers that a help-desk employee has access to the credentials of the organization network. The attacker has prior knowledge of the credentials or keys of the organization's main network after the attacker conducts some initial research.

- ***Stalker-ware***

  Stalker-ware is software that is used for cyber-stalking, as the name suggests. It is frequently placed on victims' devices without their knowledge, with the purpose of tracking all of their actions. Stalker-ware is a technology that is used for cyber-stalking, as the name implies. It's frequently placed on victims' devices without their knowledge, with the purpose of tracking all of their actions. The stalker-ware can be installed on frequently used devices like laptops, PCs, tablets, or mobile phones, without any prior knowledge, something done unintentionally or by a person who wants to stalk others [36, 37, 38, 39]. All the movements and actions of the person being stalked are monitored or measured by the stalker-ware.

8.6 FUTURE SCOPE AND CONCLUSION

Intrusion detection and intrusion prevention systems are designed to protect networks and devices from malicious intrusions and attacks from the external environment, unlike firewalls. In such cases, intrusion detection systems can be handy and assist when it comes to distinguishing between malicious traffic and regular traffic inside a network. Any network or device can be more vulnerable and easy prey for the attackers if it doesn't have a firewall, the availability of security services, or up-gradation provided by the firewalls. Intrusion detection and prevention can be an important asset for the defense against modern and trending variants of cyberattacks. But the limitation of these kinds of systems is that they operate passively. IDSs can only perform division between attacks and normal traffic and cannot help to do so in real-time.

This requires examination of the packet traffic throughout its functionality time frame and the need for a storage unit for storing the incoming and outgoing traffic patterns and attributes. Then the role of IDSs will be to distinguish between normal traffic and traffic causing problems to the network (anomaly detection). Another extension for IDSs could be if the system is able to report and block malicious sources using the source related attributes of a malicious entity.

REFERENCES

1. Kumar, B.S., Ch, T., Raju, R.S.P., Ratnakar, M., Baba, S.D. and Sudhakar, N., 2013. Intrusion detection system-types and prevention.
2. Liao, H.J., Lin, C.H.R., Lin, Y.C. and Tung, K.Y., 2013. Intrusion detection system: A comprehensive review. *Journal of Network and Computer Applications*, *36*(1), pp. 16–24.
3. Othman, S.M., Alsohybe, N.T., Ba-Alwi, F.M. and Zahary, A.T., 2018. Survey on intrusion detection system types. *International Journal of Cyber-Security and Digital Forensics*, *7*(4), pp. 444–463.

4. Hoque, M.S., Mukit, M., Bikas, M. and Naser, A., 2012. An implementation of intrusion detection system using genetic algorithm. arXiv preprintarXiv:1204.1336.
5. Wu, M. and Moon, Y.B., 2019. Intrusion detection of CYBER-PHYSICAL attacks in manufacturing systems: A review. *Volume 2B: AdvancedManufacturing.* The link educates about the variety of intrusion detection systems. https://purplesec.us/intrusion-detection-vs-intrusion-prevention-systems/.
6. Forouzan, A.B., 2007. *Data communications & networking (sie).* Tata McGraw-Hill Education.
7. Noonan, W. and Dubrawsky, I., 2006. *Firewall fundamentals.* Pearson Education.
8. Sharma, R. and Parekh, C., 2017. Firewalls: A study and its classification. *International Journal of Advanced Research in Computer Science, 8*(5).
9. Pesé, M.D., Schmidt, K. and Zweck, H., 2017. *Hardware/software co-design of an automotive embedded firewall* (No. 2017-01-1659). SAE Technical Paper.
10. Khawandi, S., Abdallah, F. and Ismail, A., A survey on image spam detection techniques. *Computer Science & Information Technology*, p. 13.
11. Idris, I., Selamat, A., Thanh Nguyen, N., Omatu, S., Krejcar, O., Kuca, K. and Penhaker, M., 2015. A combined negative selection algorithm–particle swarm optimization for an email spam detection system. *Engineering Applications of Artificial Intelligence, 39*, pp. 33–44.
12. Lallie, H.S., Shepherd, L.A., Nurse, J.R., Erola, A., Epiphaniou, G., Maple, C. and Bellekens, X., 2020. Cyber security in the age of covid-19: A timeline and analysis of cyber-crime and cyber-attacks during the pandemic. *arXiv preprint* arXiv:2006.11929.
13. Sarvi, M., Mohamadi, M. and Varjani, A.Y., 2013, August. A fuzzy expert system approach for spam detection. In *2013 13th Iranian Conference on Fuzzy Systems (IFSC)*, IEEE, pp. 1–5.
14. Rid, T. and Buchanan, B., 2015. Attributing cyber attacks. *Journal of Strategic Studies, 38*(1–2), pp. 4–37.
15. Cashell, B., Jackson, W.D., Jickling, M. and Webel, B., 2004. The economic impact of cyber-attacks. *Congressional Research Service Documents, CRS RL32331 (Washington DC), 2.*
16. Cohen, F., 1999. Simulating cyber attacks, defences, and consequences. *Computers & Security, 18*(6), pp. 479–518.
17. Li, X., Liang, X., Lu, R., Shen, X., Lin, X. and Zhu, H., 2012. Securing smart grid: Cyber attacks, countermeasures, and challenges. *IEEE Communications Magazine, 50*(8), pp. 38–45.
18. Bendovschi, A., 2015. Cyber-attacks–trends, patterns and security countermeasures. *Procedia Economics and Finance, 28*, pp. 24–31.
19. Developing Story: COVID-19 Used in Malicious Campaigns, n.d. Retrieved from https://www.trendmicro.com/vinfo/in/security/news/cybercrime-and-digital-threats/coronavirus-used-in-spam-malware-file-names-and-malicious-domains#:~:text=COVID-19 is being used, as a lure likewise increase.
20. Mahjabin, T., Xiao, Y., Sun, G. and Jiang, W., 2017. A survey of distributed denial-of-service attack, prevention, and mitigation techniques. *International Journal of Distributed Sensor Networks, 13*(12), p. 1550147717741463.
21. Irwin, L., 2020, October 12. DDoS attacks soar as organisations struggle with effects of COVID-19. Retrieved from https://www.itgovernance.eu/blog/en/ddos-attacks-soar-as-organisations-struggle-with-effects-of-covid-19#:~:text=DDoS attacks soar as organisations struggle with effects of COVID-19,-Luke Irwin 13th&text=Organisations faced two-and-a, according to a Neustar report.
22. Pressman, A., 2020, November 16. Hackers are trying to disrupt and steal COVID-19 vaccine research. Retrieved from https://fortune.com/2020/11/13/covid-vaccine- hackers-cyberattack-coronavirus-north-korea-russia/

23. Jansson, K. and von Solms, R., 2013. Phishing for phishing awareness. *Behaviour & Information Technology*, *32*(6), pp. 584–593.
24. Brewer, R., 2016. Ransomware Attacks: Detection, Prevention and Cure. *Network Security*, *2016*(9), pp. 5–9.
25. Sadgali, I., Sael, N. and Benabbou, F., 2019. Performance of machine learning techniques in the detection of financial frauds. *Procedia Computer Science*, *148*, pp. 45–54.
26. Aijaz, I. and Agarwal, P., 2020. A study on time series forecasting using hybridization of time series models and neural networks. *Recent Advances in Computer Science and Communications (Formerly: Recent Patents on Computer Science)*, *13*(5), pp. 827–832.
27. Hong, J., 2012. The state of phishing attacks. *Communications of the ACM*, *55*(1), pp. 74–81.
28. Khan, W.Z., Khan, M.K., Muhaya, F.T.B., Aalsalem, M.Y. and Chao, H.C., 2015. A comprehensive study of email spam botnet detection. *IEEE Communications Surveys & Tutorials*, *17*(4), pp. 2271–2295.
29. Salvi, M.H.U. and Kerkar, M.R.V., 2016. Ransomware: A cyber extortion. *Asian Journal for Convergence in Technology (AJCT)*, 2.
30. Axelsson, S., 2000. *Intrusion detection systems: A survey and taxonomy* (Vol. 99, pp. 1–15). Technical Report.
31. Debar, H., Dacier, M. and Wespi, A., 2000, July. A revised taxonomy for intrusion-detection systems. In *Annales des télécommunications* (Vol. 55, No. 7, pp. 361–378). Springer-Verlag.
32. Jan, S.U., Ahmed, S., Shakhov, V. and Koo, I., 2019. Toward a lightweight intrusion detection system for the internet of things. *IEEE Access*, *7*, pp. 42450–42471.
33. Hong, G., Yang, Z., Yang, S., Zhang, L., Nan, Y., Zhang, Z., Yang, M., Zhang, Y., Qian, Z. and Duan, H., 2018, October. How you get shot in the back: A systematical study about cryptojacking in the real world. In *Proceedings of the 2018 ACM SIGSAC Conference on Computer and Communications Security* (pp. 1701–1713).
34. The link is reported by Kaspersky about modern technology trending cyber-attacks, including crypto-jacking. https://www.kaspersky.com/resource-center/definitions/what-is-cryptojacking.
35. Weir, G.R., Toolan, F. and Smeed, D., 2011. The threats of social networking: Old wine in new bottles? *Information Security Technical Report*, *16*(2), pp. 38–43.
36. Shelke, P. and Badiye, A., 2013. Social networking: Its uses and abuses. *Research Journal of Forensic Sciences*, *1*(1), pp. 2–7.
37. Krombholz, K., Hobel, H., Huber, M. and Weippl, E., 2015. Advanced social engineering attacks. *Journal of Information Security and Applications*, *22*, pp. 113–122.
38. Mouton, F., Leenen, L. and Venter, H.S., 2016. Social engineering attack examples, templates and scenarios. *Computers & Security*, *59*, pp. 186–209.
39. Khoo, C., Robertson, K. and Deibert, R., 2019. Installing fear: A Canadian legal and policy analysis of using, developing, and selling smartphone spyware and stalkerware application.

9 War of Control Hijacking
Attacks and Defenses

Ragini Karwayun and Monika Sainger

CONTENTS

DOI: 10.1201/9781003229704-9

9.1 INTRODUCTION

Computer system design, implementation, operation, and maintenance all have vulnerabilities that can be exploited by attackers. A "computer system" is not just limited to hardware and software. Policies, procedures, and organization under which the hardware and software is used are also an integral part of the computer system. Vulnerabilities in any one or combination of these components can play an important role in the exploitation of the security of the system. In this chapter, we will limit our discussion to the vulnerabilities in the software, more specifically the languages that are used to build the software.

Control flow is the sequence in which every single statement, instruction, or function call of a program is executed. A control-hijacking attack exploits a program error, especially a memory corruption vulnerability, at execution runtime to sabotage the intended control flow of a program. Control-hijacking or control-flow hijacking attacks alter a code pointer that manipulates the contents of the program counter, thus gaining control of the instruction pointer. They achieve this by gaining access to prohibited memory regions that should not be accessed normally.

Software threats have seen tremendous growth in the recent past. Vulnerabilities associated with applications and software have resulted in huge losses to different organizations and individuals as a result of security attacks. Figure 9.1 illustrates a brief categorization of the seriousness of vulnerabilities identified in applications.

9.1.1 Control Flow

Control flow defines the order in which each instruction or function call of a computer program is executed. Code written using an imperative programming language has a certain control flow; the statements are used to change a program's state. However, the declarative programming paradigm does not explicitly describe the control flow of a code; it rather defines the structure and elements of computer programs expressing the logic of computation.

In imperative programming languages, a control-flow statement is a statement that makes the program move forward by choosing between any one path from two

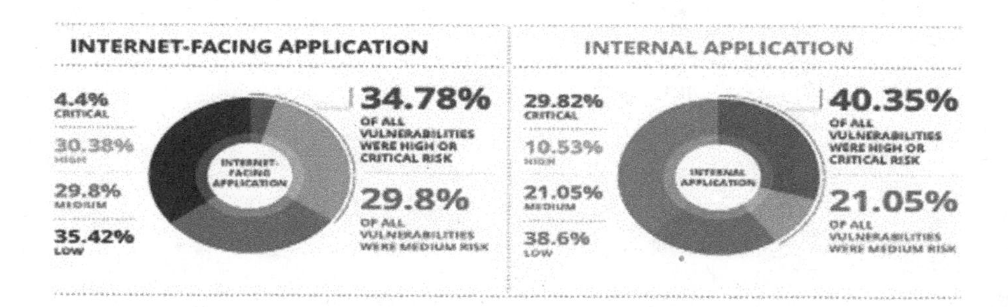

FIGURE 9.1 Vulnerability Statistics Report. *Source: EdgeScan 2020.*

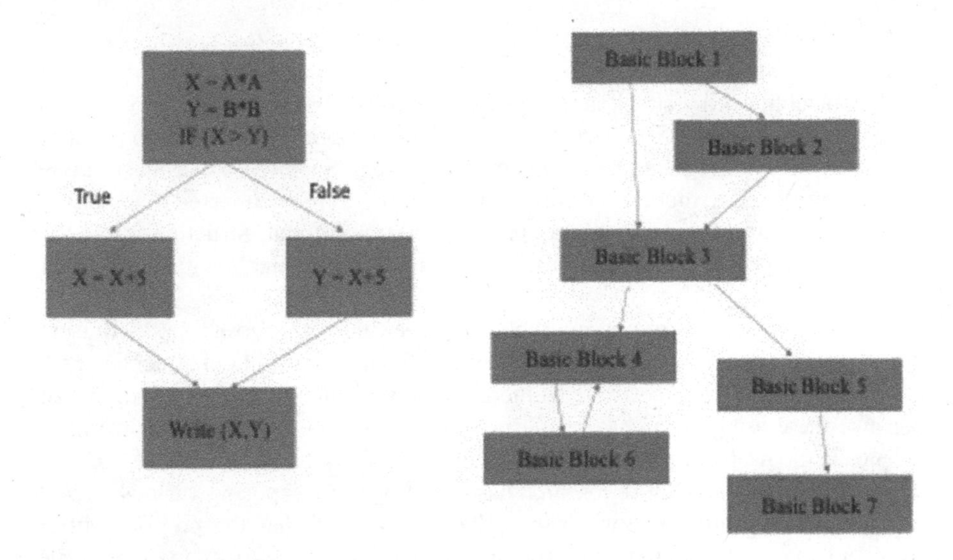

FIGURE 9.2 Control-flow graphs.

or more options available. Low-level implementations like interrupts and signals can modify the flow of control in the same way as a subroutine or a function. These events generally occur as a response to some external asynchronous event, rather than as a result of executing an in-line control-flow statement.

In lower-level languages, like machine or assembly language, control-flow instructions are executed by changing the contents of the program counter register. For some Central Processing Units (CPUs), there are only a few control-flow instructions available, such as BRANCH (conditional or unconditional) instructions, also referred to as JUMP.

9.1.2 Security Concerns

Current software exploits vulnerabilities that sabotage machine-code execution. Quite often computers are subjected to external attacks with the sole objective to control how software behaves. Attackers send malicious data over a regular communication channel and, once it is resident in program memory, an already-existing software flaw is manipulated to perform the attack. Among the most common are control-flow attacks, including buffer overflows, code-reuse attacks, return-to-libc, code gadgets, and Linux rootkits, and many others.

9.2 CONTROL HIJACKING

Control-flow hijacking manipulates memory corruption vulnerabilities to alter the normal program execution away from the pre-defined control flow. Memory corruption errors in C/C++/JAVA programs are the most common source of security vulnerabilities in the present scenario [1]. Even though during the last decade, several defense mechanisms have been developed, code reuse and control-flow hijacking are still challenging affairs [2].

A control-hijacking attack is performed by altering the intended control flow by overwriting some of the data structures in a victim program. This leads to hijacking the control of the program and the underlying system. Such types of data structures that can be altered to modify the control flow of a program are referred to as *control-sensitive* data structures. Some examples of such types of data structures are return address, function pointer, global offset table, and virtual functions table pointer in C++, etc.

Attackers can perform any operation after getting control over the victim program. Control-hijacking attacks are believed to be the most critical attacks because they are able to exploit software vulnerabilities without any user intervention and because these are used as basic building blocks by malicious computer worms to propagate themselves on multiple machines [3].

A considerable amount of research has been done into exposing various possible control-hijacking attacks. Some approaches identify whether the program suffers from buffer overflow using program analysis techniques [4–6]. Some approaches use program transformation techniques [7–9] to convert applications such that the new versions can either prevent control-sensitive data structures from being altered at runtime [10], or [4, 9, 11, 12] have suggested schemes that can detect control-hijacking attacks. Whereas there are some approaches that have developed operating system mechanisms negating the possibility of executing code that is injected into the victim program [11, 13].

Most of the proposed approaches were only able to identify a program under a control-hijacking attack. The only solution was to terminate the victim program instead of trying to repair the effects of the hijacking. The victim program could be restarted if required. But this solution resulted in a denial-of-service attack, even if further instances of the attack were prevented.

In [13], researchers proposed a system that enables an application to detect a control-hijacking attack when its control-sensitive data structures are tampered with. It prevents further occurrences of such types of attacks, and it also has a repairing mechanism that erases all the side effects of the attack packets.

There are various ways in which attackers can perform control-hijacking attacks. Control-hijacking attacks exploit a memory corruption vulnerability by altering the pre-defined control flow of a program at application runtime. Attackers try to change the control flow by changing the content of a code pointer (i.e., the value that defines the content of the program counter) or obtain control of the instruction pointer or violet protected memory region.

The first step in a control-hijacking attack is to alter the value of a pointer such that it points to the malicious code of the attacker. Next, code pointers are updated so that the intended control flow can be compromised. This can be achieved by performing buffer overflow. Finally, the malicious code is executed, resulting in obtaining sensitive and secret information or the attacker gaining unauthorized high-access privileges. A typical control-hijacking attack takes place as follows:

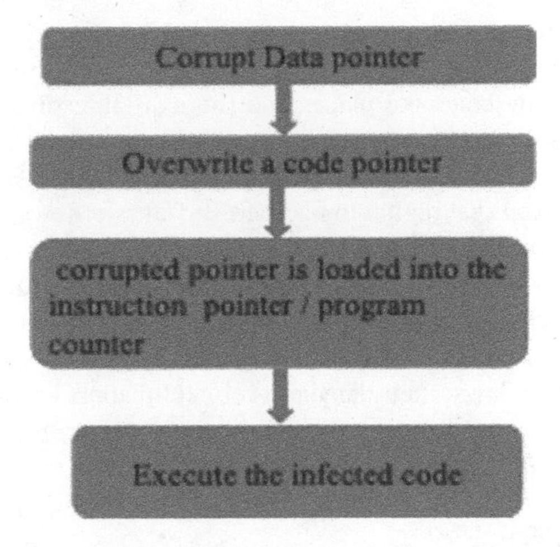

FIGURE 9.3 Process flow of control hijacking.

Control-hijacking attacks can be classified into code-injection and code-reuse attacks. In a code-injection attack, a new node is added to the CFG to execute arbitrary malicious code, whereas a new path is added to the CFG in a code-reuse attack.

In the following section, we will discuss different types of control-hijacking attacks.

9.2.1 Buffer-Overflow Attacks

In 1988, the earliest internet era, a very peculiar type of malware known as the Morris Worm infected nearly 10% of the internet in only two days, costing approximately $100,000 and $10 million in damage, as per the records of the Government Accountability Office. The Morris Worm is commonly known as the buffer-overflow attack. Buffer overflow is a software coding anomaly or vulnerability that an attacker can exploit to gain unauthorized access to any system.

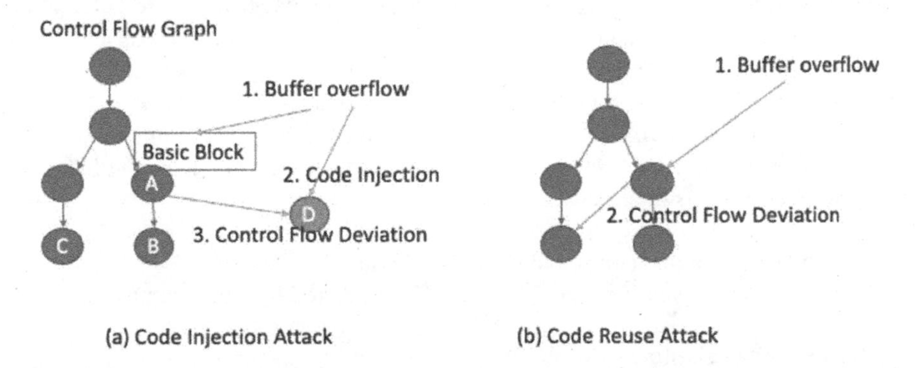

FIGURE 9.4 Code injection attack and code reuse attack. Source: Lecture: Code-Reuse Attacks and Defenses, Lucas Davi, Winter School on Binary Analysis, 2017.

Even though this is one of the most common software vulnerabilities, it is not fully controlled. This is due to the fact that there are multiple different ways in which buffer overflow can take place, and the available prevention techniques are not fool proof and error-free.

Buffers are sequential memory storage locations that hold data for the time being as it is transferred from one region to another. Buffer overflows, also referred to as buffer overruns, happen when the amount of data transferred in the buffer overflows its capacity. The data that overflows into contiguous memory locations can corrupt the content of the memory and can be thus used to perform malicious activities. To effectively manage and control buffer-overflow vulnerabilities, it is necessary to understand buffer overflows, their dangerous effect on applications, and the ways in which attackers can exploit them.

Regardless of the fact that it is an extremely difficult task to discover buffer overflows and exploit them, attackers have been able to identify buffer overflows in a wide variety of products.

In a classic case of a buffer-overflow attack, the attacker tries to store data in a stack buffer that has less size than the data. As a result, the information on the call stack as well as the function's return pointer is overwritten by this overflow data. The value of the return pointer now holds overflow data, and hence after the function returns, the control is transferred to the malicious code included in the attacker's data.

There are a wide variety of buffer-overflow attacks that use distinct strategies and target different code snippets. Some of the most well-known buffer-overflow attacks are:

- Stack overflow
- Heap overflow
- Return-to-libc
- Return-oriented programming
- Heap spray attacks

As compared to other languages, C and C++ coding languages are more vulnerable to buffer overflow. They do not have any built-in protection against gaining access or overwriting data in their memory. The most popular operating systems like Windows, Mac OS, and Linux are written in one or both of these languages.

Many languages like JAVA and C# contain inherent features that help to lower the chances of buffer overflow but cannot prevent it completely. Code that is vulnerable to buffer overflow has the following properties:

- Its behavior can be controlled by external data.
- Its complex behavior cannot be accurately predicted by a programmer.

Buffer-Overflow Exploits

Attackers must be aware of the architecture and operating system of the target machine. For example, if an attacker is aware of the program's memory blueprint, they may purposely input extra data that is more than the buffer size and thus overflows into adjacent memory locations. This will enable them to store malicious code that in turn enables them to take control of the program.

Attackers may use a buffer overflow to overwrite an application's execution stack and seize the control of the machine by executing arbitrary code. Buffer overflows can exist in both the application servers and web servers, particularly web applications using graphic libraries.

9.2.2 Stack-Based Buffer Overflows

The stack is a very popular data structure that acts as an array of data items with two key operations: *push*, which inserts an item to the array, and *pop*, which deletes the topmost based on the LIFO (Last In First Out) method.

Each function has an associated local memory that holds incoming parameters during function calls, local and temporary variables. This memory area is referred to as a stack frame. A frame pointer register (ebp) contains the base address of the function's frame. The value of the stack pointer register (esp) changes during the execution of a function as items are pushed onto or popped off the stack (such as pushing parameters in preparation for calling another function). The frame pointer remains the same throughout the function [14].

To perform a buffer overflow, one can write arbitrary data onto the stack. This means that the return address of a function can be changed and also overwritten on the adjacent memory area beyond the return address – into the local variables of previous functions.

Stack overflow is the most common type of buffer-overflow attack. A stack-based attack happens when an attacker sends data consisting of malicious code to an application, where the data is stored in a stack buffer. Thus stack is overwritten with this data, including its return pointer, which in turn transfers the control of the program to the attacker.

The following program snippet has a function with a simple buffer-overflow coding error. The function "example" copies a given string without any checks by using strcpy() instead of strncpy(). The contents of (l_str[]) are copied into buff[] until it finds a null character in the string. As buff[] is much smaller than *str the extra 240

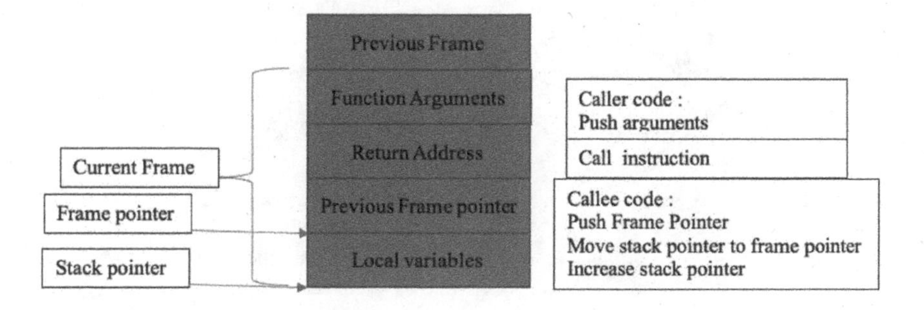

FIGURE 9.5 General stack structure.

bytes after the buffer of 16 bytes in the stack will be overwritten, including SFP, RET, and *str. The Hex character value of A is 0x41.

```
void example(char *str)
    {
        char buf[16];
        strcpy(buf,str);
    }
int main(void)
    {
        char l_str[256];
        int i;
        for(i=0; i<255; i++) l_str[i]='A';
        example(l_str); return0;}
     }
```

That makes the return address 0x41414141. This is beyond the process address space. Therefore, when the function attempts to read the return address, a segmentation violation happens. But this overflow enables an attacker to change the return address of a function. As a result, the flow of execution of the program can be changed. In most cases, this is used to execute the shell.

9.2.3 Heap-Based Buffer Overflows

A heap-based attack involves overflowing a program's memory space further than the memory that is used for runtime operations. Heap overflows are more complex to perform than stack-based buffer overflows. Since heap-based attacks try to overwrite information in the heap, and since heaps are used to contain dynamically allocated memory information, return addresses are not saved in the heap, and the return address of a function cannot be overwritten using a shellcode. Therefore, different techniques must be used by an attacker to acquire control of the execution flow. An attacker could overwrite the function pointers stored in these memory regions can be overwritten by an attacker, but quite often these are not present in memory.

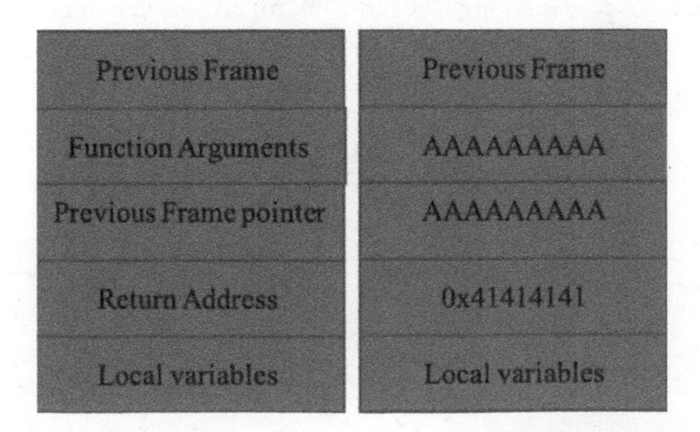

FIGURE 9.6 Stack after overflow attack.

Heap-based overflow attacks involve overwriting the dynamically allocated memory information. Memory is allocated in chunks. These chunks store the memory management information alongside the actual data. Dynamic memory allocators such as dlmalloc and ptmalloc can be attacked by overwriting the chunk info.

The chunks of memory administered by Doug Lea's Malloc (dlmalloc) consist of the size of information fields that precedes and succeeds the chunk [15]. These chunks have a variable size as various allocation routines are used, such as malloc, free, or realloc. The free chunks are managed as a doubly linked list sorted in nondecreasing order. The allocation algorithm used is an improvised version of the best-fit bin-packing heuristic. When dlmalloc needs to assign a free chunk, the first chunk of that size is allocated.

Let us take an example of four memory chunks A, B, C, and D. Chunk A is allocated, and B, C, and D are free chunks. These chunks are placed at arbitrary locations in the memory such that C is adjacent to A.

Chunk B, C, and D are connected in a doubly linked list in that order. The presence of the boundary tag between each chunk of memory provides an opportunity for the attacker to exploit heap mismanagement. An attacker exploits dlmalloc by processing these boundary tags, thus enabling them to execute arbitrary code. So in Figure 9.7, the return address will be overwritten with a pointer value to the code that will overflow into the place where the forward pointer is stored and will execute the injected code. In this way, random memory locations can be overwritten. Deallocating the memory twice can also result in double-free vulnerability.

There are multiple ways in which heap-overflow attacks can happen. For example, the attacker could overflow a dynamically allocated buffer and overwrite the next successive boundary tag (Netscape browsers exploit) or underflow such a buffer and overwrites the boundary tag that precedes the buffer (Secure Locate exploit) or forces the vulnerable program to perform an incorrect free call (LBNL traceroute exploit) or multiple frees or overwrite a single byte of a boundary tag with a NULL byte (Sudo exploit).

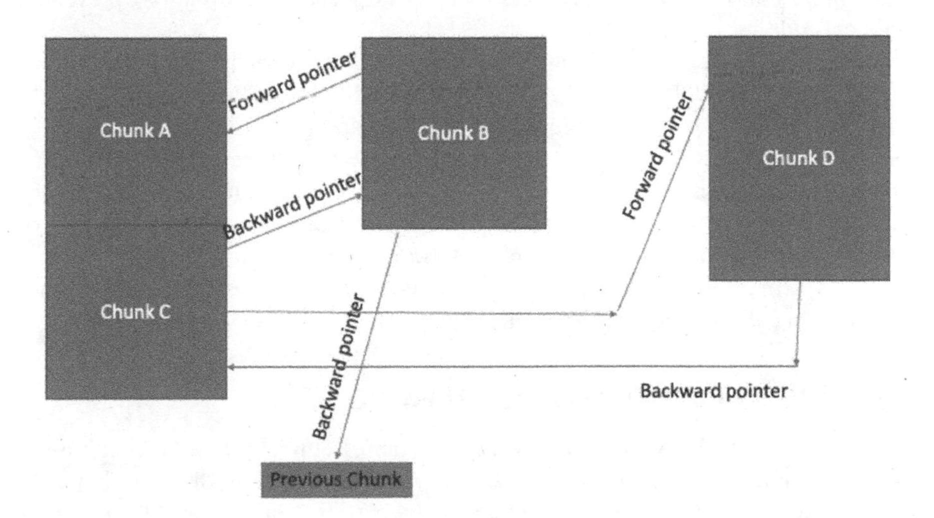

FIGURE 9.7 Heap containing used and free chunks.

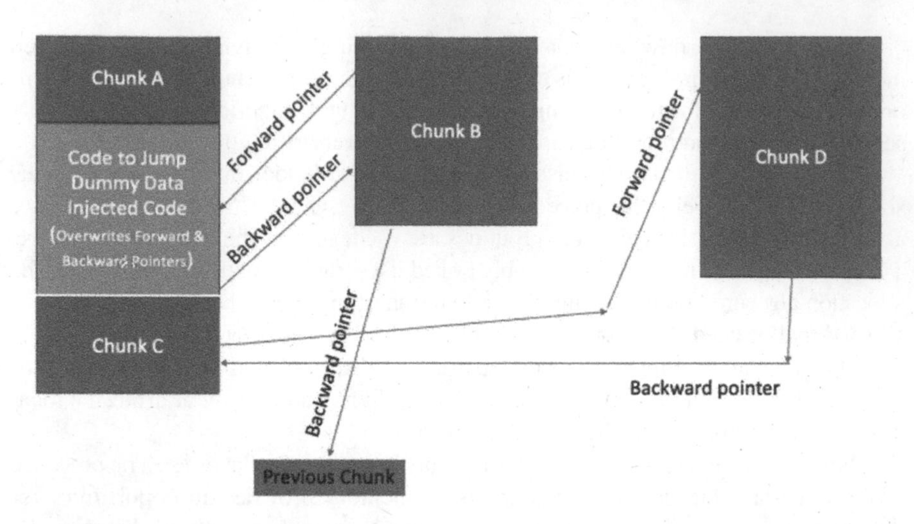

FIGURE 9.8 Heap-based buffer overflow.

9.2.4 Return to libc

Attackers exploit the buffer-overflow vulnerability by overflowing the buffer with a malicious code, and then the code injection is used to change the control flow of the victim program to jump to the location in the stack where the shellcode is stored. Quite often, attackers are not able to jump to the exact location of the shellcode, so they use NOP (no operation) slides to reach the desired location. Many researchers have proposed hardware and software mechanisms to protect against these attacks. One such approach uses a hardware memory protection method that makes the memory page either writable or executable but not at the same time (referred to as $W \oplus X$ protection). This method, referred to as DEP (Data Execution Prevention), prevented classical code injection attacks [16].

But the DEP scheme is not fool-proof; return-to-libc, a variant of buffer-overflow, can outsmart DEP. This attack does not require an executable stack, or a shellcode to perform an exploit. Instead, it makes the vulnerable program jump to some already loaded code present in memory, such as some C library functions like system(). Libraries contain definitions of all functions. So, address to a function in a C library can overwrite a return address and then change the arguments and saved return addresses. In this way, a fake function stack frame can be created.

To perform a libc attack, we try to get the address of the system() call (or other such system calls) by multiple trials and then pass "bin/sh" as an argument to it.

9.3 RETURN-ORIENTED PROGRAMMING

As the war flares up between attackers and defenders, Return-Oriented Programming (ROP) is another strong weapon in the arms race. The war of control hijacking started with the exploits of language vulnerabilities in coding; OS designers responded with DEP, and as a counterattack ROP was used as a tool to circumvent

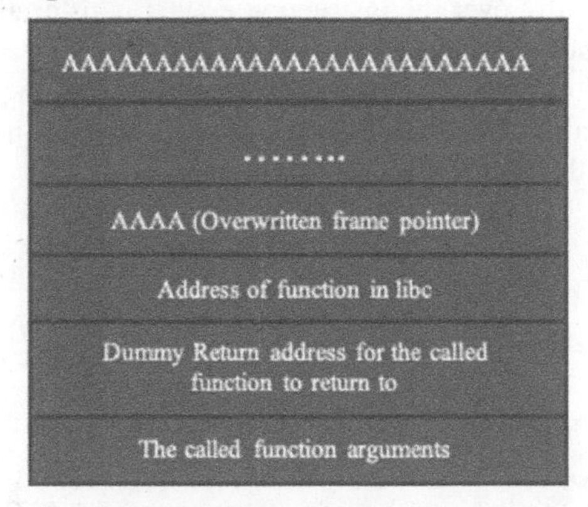

FIGURE 9.9 Stack after a return-to-libc attack.

DEP. The technique of using simple and harmless code to perform malicious activities was introduced in 2007 by Hovav Shacham, which was later referred to as Return-Oriented Programming [15]. Attackers can use ROP to bypass the protection provided by the OS using DEP. Using ROP, an attacker after diverting the control flow can force the victim program to behave in an arbitrary manner. Multiple short instruction sequences, each having "return" as the last instruction, are chained together in an ROP program. These instructions are already present in a program's address space. Return-oriented programming eludes the protection provided by the W⊕X operation.

ROP is named after the RETN assembly instruction, which is responsible for maintaining the control flow of the program in cases of function calls. RETN takes the saved instruction pointer value from the stack and updates the stack pointer, and the instruction pointer registers %esp and %eip. If an attacker is able to perform a buffer-overflow exploit on the target system, it can update the appropriate place in the memory with its own malicious code. As a result as soon as the execution flow reaches RETN instruction, the instruction pointer is referred and the execution continues from the pointed memory cell.

In ROP each instruction sequence ending in "RETN" is called a gadget. If attackers are able to identify various gadgets, they can initiate a chain reaction of return statements to execute arbitrary malicious code. The first RETN instruction or gadget in the chain is invoked when its address is placed on the victim stack of the buffer-overflow attack. After the gadget is executed, the attackers get control over the program flow because every gadget ends with a RETN. By changing the next return address on the stack, the control flow can be redirected to another gadget until the attacker wishes to end the code execution. Because gadget chaining is particularly a series of jumps to pieces of existing code, it is considered to be an extremely complex procedure to exploit it for malicious purposes. However, the ROP technique uses code misplacement to falsify new instructions from code already loaded in memory.

To perform ROP, we first need to perform a stack buffer overflow on a variable that can give us control over the instruction pointer register after the main returns. Then we can pass our arguments to the system call. Hence, we can get control of the stack if we have control of the arguments passed on to the stack.

So, when the main returns, the fake stack should behave like there was a normal system function call. After a function has been called, a stack looks like:

```
...                                         // More arguments
0xffff0008: 0x00000002                      // Argument 2
0xffff0004: 0x00000001                      // Argument 1
ESP ->0xffff0000: 0x080484d0                // Return address
```

So, main's stack frame needs to look like this:

```
0xffff0008: 0xdeadbeef         // system argument 1
0xffff0004: 0xdeadbeef         // return address for system
ESP ->0xffff0000: 0x08048450   // return address for main
                                  (system's PLT entry)
```

Then when the main returns, it will hop into the system's procedure linkage table entry, and the stack looks normal as if the system had been called for the first time.

9.4 HEAP SPRAY ATTACKS

Heap spraying was first documented by Skylined in 2001. Skylined performed an IFRAME tag buffer exploit for Internet Explorer in 2004 using this technique. After many years, it is still a popular payload delivery technique in browser exploits. Many efforts have been made for the detection and prevention of heap sprays, but the concept still works. Heap spraying is a payload delivery technique used in buffer-overflow exploits to assist the execution of arbitrary code. The goal is to place a shellcode at an expected address in the victim application. In this way, some vulnerability can be used to execute this shellcode. This technique is implemented by part of an exploit's source code referred to as the heap spray.

As heaps are used for dynamic memory allocation, implementing dynamic memory managers is a challenging task. Heap fragmentation adds to the difficulty level. To perform a heap spray attack, we need to have technical expertise to be able to allocate free chunks of memory in the heap before gaining control over the instruction pointer. Heap spraying involves allocating multiple copies of the exploit code to increase the chances of any one copy to find space in the heap. The idea is to have one random program jump to reach any one copy of the exploit code present in the heap to start the attack. Hence, we can easily say that the heap spray is itself not an exploit, but it provides easier means to exploit an existing vulnerability.

A heap-spray attack consists of two stages – memory allocation stage and execution stage. In the allocation stage, lots of fixed-size chunks in the memory are allocated to the same content. In the execution phase, one of the several chunks loaded gains control over the memory.

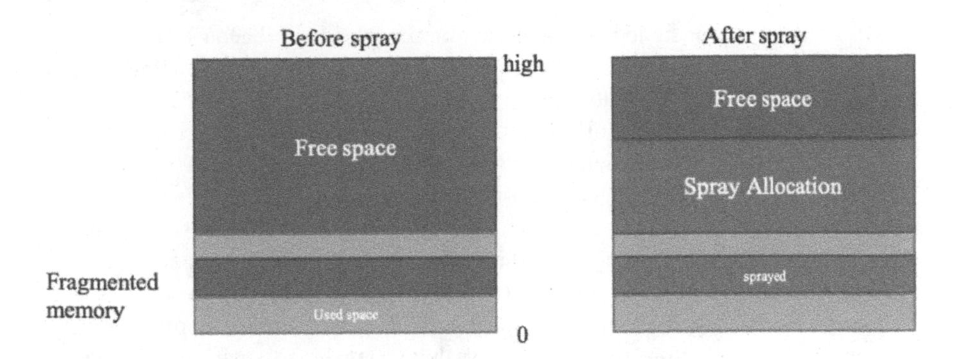

FIGURE 9.10 Heap spraying example.

Web browsers are the easiest to compromise by using JavaScript. Any application that supports scripting languages is a prospective victim of a heap spraying attack.

Some popular buffer-overflow attacks: Buffer overflows are still the most common security vulnerability, and they have been responsible for various data breaches [17]. Some of the most popular are Morris Worm (1988–1990), SQL Slammer (2003), Adobe Flash Player (2016), and WhatsApp VoIP (2019).

9.5 INTEGER OVERFLOW ATTACKS

The integer overflow occurs when an arithmetic operation tries to generate a numeric value that is larger than the storage space allocated to store that value. It has been a very common problem for a very long time, but now integer overflow vulnerabilities are used by hackers. There is a race between the number of integer overflow vulnerabilities exploited and the detection methods used for integer overflows, both growing at a very rapid pace.

Each integer type in C has a fixed minimum and maximum value that depends on the type's machine representation – 2's complement vs. 1's complement, signed or unsigned, and 16-bits or 32-bits. At the application level, programmers often do not take into consideration the maximum and minimum values an integer can take, resulting in integer vulnerabilities. Integer vulnerabilities can be classified into four categories: overflows, underflows, truncations, and sign conversion errors [18].

Overflows – An integer overflow is caused at runtime when the result of an integer expression is greater than the maximum value its respective type can represent. For example, an unsigned 16-bit integer value that may store any value from 0 to 65,535, or a signed integer that can have any value from –32,768 to 32,767. So, during an arithmetic operation, if the results require more than the allocated space (adding 1 to the maximum value), most compilers will ignore the overflow and store unexpected output or error. Integer overflow exploit can be used as a tool to perform various attacks such as buffer overflow that results in executing malicious programs or privilege escalation. The C Standard [ISO/IEC 9899:2011], states: "A computation comprising of unsigned operands can never overflow" because the result can be reduced modulo the result type's width." However, overflows are currently the most common integer vulnerability, accounting for more than half of the CVE vulnerabilities surveyed [18], reflecting the fact that most programmers do not understand or anticipate the C Standard.

Underflow – An integer underflow also occurs at execution time when the result of an integer expression is smaller than the minimum value, thus rounding off to the maximum integer for the type. For example, if one is subtracted from an unsigned variable with a current value of zero, this would result in a value of negative one. But as unsigned variables cannot store negative numbers, it will store the maximum value an unsigned variable can hold. Since underflows normally occur only with subtraction, they are rarer than overflows.

Signedness Error – A signedness error occurs as a result of misinterpretation of signed and unsigned integers. In 2's -complement representation, the sign bit can be interpreted as the most significant bit (MSB) or vice versa; hence -1 and $2-1$ are misinterpreted by each other on 32-bit machines. This also accounts for several vulnerabilities in the survey conducted [18].

Truncation – This happens when an integer with a larger width is assigned to an integer with a smaller width. For example, assigning an int to a short intger ignores the leading bits of the int value, which results in potential information loss.

The 2020 CWE most dangerous software vulnerabilities placed integer overflow in 11th place. This weakness is considered very critical because it can be very easily located and exploited. Through this exploitation, the attacker is able to take over the system completely and can then perform any type of malicious activity like data theft, data exfiltration, or DoS. This error is usually introduced in the implementation of the software lifecycle. There are more than 1,113 vulnerabilities in the Common Vulnerability Exposure database that are associated with integer overflow attacks.

9.6 FORMAT STRING VULNERABILITIES

A format string attack happens when an application's input data is processed as a command or the input data is not validated effectively. This enables code execution, reading content of the stack, or causing segmentation faults by the attackers. This could trigger new actions that threaten the security and stability of the system.

Almost every programming language uses format strings for inserting values into a text string. One of the foremost functions used in the C programming language is printf(), or "print formatted." Format functions are extremely convenient and handy for generating readable output as they provide automatic type conversions. However, if one is not careful enough, vulnerability of the printf() format strings can be exploited to perform a variety of attacks. Actually, the entire family of the printf() format functions including fprintf(), snprintf(), vpprintf(), etc., are all vulnerable.

A major new category of vulnerabilities called "format bugs" was discovered in June 2000 when a vulnerability was detected in WU-FTP that was similar to buffer overflow [19]. It was observed that it is dangerous to permit potentially hostile input to be passed directly as the format string for calls to format functions. As these functions allow automatic type conversion, including the "%" directive in the format string, it enables the attackers to perform unexpected or malicious activities.

Some common format parameters that can be used in a format string attack are:

Parameter	Passed as
%p	Reference
%d	Decimal value
%x	Hexa decimal value
%s	Reference
%n	Reference

Let us consider a line of code – printf("%s," argv[1]), which is a simple and safe instruction. But if we rewrite this instruction as printf(argv[1]), it becomes vulnerable. For example, if we pass "string attack %s%s%s%s%s%s" to the first line it will simply print: "string attack %s%s%s%s%s%s" but executing the second line will result in the system interpreting every %s as a reference to a string starting from the buffer location, probably on the stack. The format string controls the behavior of the format function. The function retrieves the requested parameters from the stack. The following situations may arise :

- There may be a mismatch between the format string and the actual parameters. For example, in the statement printf("Value of A=%d, Value of B=%d, Value of C=%d and address of c=%x\n",a,b,c), the format string asks for four arguments whereas only three are supplied. The compiler will not be able to identify the mismatch because of the flexibility provided in the printf() function. The function will fetch four data items from the stack, and in case of mismatch it will try to read some information from the stack that is out of the bound of this function call.
- Using "%s%s%s%s%s" each %s will fetch a number from the stack; this number will be treated as an address and the contents of the memory pointed to by this address will be printed. If the address space referred to is protected, then this will result in the crashing of the program.
- Printf("%o8x %08x….") to print the contents of the stack.
- Using printf("%s"), the contents of any location of the memory can be viewed.
- The function string parameter %n counts the number of characters written so far. We can use printf with %n to overwrite any memory location. This can be used to overwrite any return address or important control flags that

control privilege access. To write at an address value of 1,000, we simply need to use padding of 1,000 dummy characters.

9.7 DEFENSES AGAINST CONTROL HIJACKING

The presence of memory corruption bugs in many traditional programming languages provides opportunities for attackers to hijack the control flow of the program. For more than three decades, applications have been affected by this problem, and multiple solutions have been proposed. Even then, control-hijacking attacks continue to be a serious threat.

A variety of defense mechanisms have been proposed to mitigate control-flow hijacking attacks. These mechanisms can be categorized into two classes – platform defenses and runtime defenses. The applicability of these defense mechanisms depends on how a particular technique balances tradeoff security with the performance overhead.

9.7.1 Platform Defenses

A large number of concurrent applications are supported by most modern operating systems. These applications may be susceptible to attacks, or perform malicious activities themselves. To protect itself, these operating systems are designed as reference monitors, and there is a strict separation between access control and resource management. The OS kernel strictly isolates kernel software and application software at runtime. Control-hijacking attacks try to corrupt memory and gain control of the system with high privileges. Some of the common mitigation techniques used for preventing these attacks are discussed below:

9.7.2 Fix Bugs

Various audit software can be used to test the software for vulnerabilities. Several automated tools are available in the market like Codacy, SonarQube, Coverity, PREfix, PREfast, and many more. Codacy is an automated tool used for code review. It supports more than 40 programming languages like Scala, Java, Ruby, JavaScript, PHP, Python, CoffeeScript, and CSS. SonarQube provides continuous improvement by giving a detailed report about the quality of the source code and highlights the issues found. Coverity performs static analysis of all the possible paths of execution through source code, and detects vulnerabilities caused by the conjunction of independently correct statements. The PREfix tool representatively executes chosen paths through a C/C++program, and during this process it looks for multiple low-level programming errors, including NULL pointer dereferences, the use of uninitialized memory, double freeing of resources, etc. PREfast analysis is inexpensive, and uses pattern matching in the syntax tree of the C/C++program to find naive programming mistakes. Other analyses are centered on local dataflow analyses to find uninitialized use of variables, NULL pointer dereferences, etc. [20]. But the major problem with these tools is that they are expansive and exclusively designed for specific software.

9.7.3 Rewriting Software

Another approach used for preventing these attacks includes rewriting the entire software using type-safe languages like JAVA. The programmers should use standard library functions that take buffer sizes into consideration. Modern versions of gcc and visual studio give a warning when a user program tries to use unsafe functions like gets(). In general, one should not ignore compiler warnings; instead, these warnings should be treated like errors. The study in [21] used a binary rewriting approach to strengthen existing Win32 binary programs with a RAD (return address defense) mechanism, which uses a redundant copy of the return address on the stack to protect its integrity.

9.7.4 Non-Executable Memory (NEP)

The idea is to categorize the memory range into either writable or executable (but not both); this prevents the execution of code injected as data into the system. It prevents code execution by marking stack and heap as non-executable. To be precise, memory regions identified as "data-only" are never treated as instructions, thus making code injection ineffective. This security policy is enforced in hardware using NX-bit on AMD Athlon 64 or XD-bit on Intel P4 Prescott. AMD names this protection feature as Enhanced Virus Protection. The NX-bit, which is used to represent *Never eXecute*, is a tool used to separately differentiate between areas of memory reserved for storage of data or for storage of processor instructions (or *code*). An operating system may identify specific areas of memory as non-executable. Any code belonging to this marked memory area will not be executed by the processor. NX-bit is used in every page table entry and is set to "1" if we wish to make the data residing on this page un-executable.

Intel defines the NEP region using XD-bit, which means *eXecute Disable*. The ARM architecture uses XN for *eXecute Never*, which was introduced in ARM v6. Linux via its PaX project implements NEP. PaX is a Linux kernel patch that specifies the least-privilege protections for pages in the memory. The least-privilege approach enables restricting the set of operations that a computer program is allowed to execute. The first version of PaX was released in 2000. PaX declares data memory as non-executable and memory where program instructions are stored as non-writable. This can help prevent some buffer-overflow security exploits by preventing the execution of unintended code. Windows has used this approach since XP SP2 and refers to it as DEP. DEP by default marks all memory locations in a process as non-executable unless the location is explicitly marked executable. If attackers try to insert and run code from NEP locations, DEP intercepts them and prevents these attacks by raising an exception. Visual Studio: /NXCompat[:NO] Indicates that an executable was tested to be compatible with the Windows Data Execution Prevention feature. The major limitation of NEP is that it cannot be used in every case. There are some applications that need an executable heap. Moreover, this approach does not offer a defense against Return-Oriented Programming exploits.

9.7.5 Address Space Layout Randomization (ASLR)

ASLR is a very popular and widely used protection technique that prevents various exploits by randomizing the memory address space layout of processes. Unlike source code analysis tools, discussed in the previous section, which focus on providing security by getting rid of vulnerabilities from the system, ASLR tries to make the process of exploitation more difficult. The security provided by ASLR depends on multiple factors [22], such as the predictability of randomization of memory layout, the effect random memory addresses have on the exploitation technique and number of attempts an attacker can try. ASLR degenerates code execution attacks into denial-of-service attacks by crashing the application. The ASLR has multiple implementations, each having its own improvisation in performance and security coverage between them.

ASLR involves the random relocation of critical parts such as the base address of the stack and the heap of the program memory, and the base of the dynamic link libraries. Since many code injection exploits need to know the absolute memory offsets and expectable memory layouts, relocation of the offsets of critical addresses would result in the crashing of the program instead of the execution of arbitrary malicious code by the attacker. The success of many attacks, especially zero-day exploits, depends on the hacker's ability to know or guess the location of processes and functions in the memory. ASLR puts critical address targets in random locations. Target application will break down if an attacker attempts to exploit an incorrect address space location, thereby stalling the attack and forewarning the system. ASLR was first introduced by the PaX project as a Linux patch in 2001, and Windows OS Vista from 2007 onwards had ASLR integrated into the system. Mac OS X 10.5 Leopard included ASLR, and both Google Android and Apple iOS started using ASLR in 2011.

To perform buffer-overflow attacks, an attacker needs to be aware of the memory layout of the program under attack. The process of finding the memory layout is a complex trial and error procedure. After this, the attacker needs to find a suitable place to inject the carefully designed malicious payload. ASLR works in conjunction with the virtual memory management to implement randomization of memory address space. Address of critical memory components like stack, heap, and dynamic libraries is changed every time the program is executed. As the address changes every time, attackers are not able to get the target address. Initially, it was required that applications are compiled with ASLR support; now this has become a default step. Windows 7 permitted 8 bits of randomness for DLL. It mapped a 64K page onto a 16 MB region. This 8-bit randomness resulted in 256 more possible address space locations. As a result, attackers only had a 1 in 256 chance of getting the hold of the correct location to execute code. Windows 8 allowed 24 bits of randomness on 64-bit processors making 1 in 224 chance of finding the correct location.

The factors that define ASLR implementation are – when, what, and how [22]. The first attribute defines the frequency of randomization, the second determines which objects are randomized, and the last one defines the reason and the extent to which the objects are randomized.

- The "when" attribute defines when the randomness for the object is generated.
  - per-ex – the random addresses are generated every time a new image is loaded in memory. Linux ASLR randomizes all objects per-exec.
  - per-boot – randomization is done at the time of booting the system. OS X performs randomization of libraries per-boot.
  - per-deployment – The application is randomized at the installation time.
  - per-fork – The randomization takes place every time a new process is created (forked).
  - per-object – The object is randomized when it is created.
- The "what" parameter defines the object that is to be randomized. ASLR is considered to be broken even if one single object is not randomized.
- The "how" attribute.
  - Partial VM – A sub-range of the virtual memory space is used to map the object.
  - Full VM – The full virtual memory space is used to map the object.
  - Isolated-object – The object is randomized independently from any other.
  - Correlated-object – The object is randomized with respect to another.
  - Sub-page – Page offset bits are randomized.
  - Bit-slicing – Different slices of the address are randomized at different times.

9.7.6 Syscall Randomization

During a code injection attack, system calls are used by the injected code to perform its malicious actions. By creating randomized system calls of the target process, an attacker, without knowing the randomization key, injects code that is different from the associated de-randomized module. Thus the injected code would not be able to execute as it is not able to call the calling system calls correctly.

9.7.7 Instruction Set Randomization (ISR)

Attackers must be aware of the instruction set of the target machine in order to have the injected code perform the desired effect. Hence, a common technique for defending against code injection attacks is to make the instruction set ambiguous to the attacker. Instruction Set Randomization (ISR) is a mitigation technique that achieves this by randomly varying the instructions used by a host machine. In this way, ISR defended all code injection attacks by changing the instruction set dynamically. But not all control-hijacking attacks can be defended by ISR. Attacks that do not use instruction sets like return-to-libc cannot be defended by ISR.

9.8 RUNTIME DEFENSES

Runtime defense approaches use stackguard canaries, libsafe, control-flow integrity, and control point integrity, and many more. The major difference lies in how each technique balances security with the performance overhead. We will be discussing each mechanism in detail.

9.8.1 Stackguard

Stackguard is an extension of the standard GNU C compiler that makes the executable code produced by the compiler better, such that it is able to detect as well as foil the stack-based buffer-overflow attacks. The normal functions of the program are oblivious of this effect. StackGuard-enabled programs specify the way the return address of a function is mentioned while it is still active. If the attacker is not able to change the return address, then there is no other way of invoking the injected attack code, and thus the attack is prevented. StackGuard uses two approaches to prevent changes to active return addresses. In the first approach, it detects the change of the return address prior to the return of the function, and in the second alternative it completely prevents the write to the return address. Discovering changes in the return address is a more effective and portable technique, while better security is provided by the prevention approach. StackGuard uses both approaches and quickly switches between the two as the need arises.

The idea is to insert a "guard" value referred to as a "canary" precedes the return address in the stack; if buffer-overflow happens and it tries to overwrite the return address, it overwrites the canary before the return address, and the system checks the changes canary before using the return address [23]. Stackguard places "canaries" (special bit patterns) in between the local variables and the return address, which act as footprints on the sand. Their integrity is verified prior to function return. If there is an overflow, it has to overwrite on the canary before overwriting on the return address. Compilers are altered to put canaries into stack frames for every function call.

The canary defense mechanism is sufficient to prevent most buffer-overflow attacks. However, it is not very difficult to write attacks that can outsmart Stackguard. Attackers can overcome the canary method of detecting overflows by either skipping over the canary word or by guessing the canary correctly. Moreover, Stackguard cannot prevent heap-based overflow attacks. To overcome some of the problems, researchers have proposed using different types of canaries

- Random Canary – Chosen randomly at the time of execution of the program. A canary string is inserted into every stack frame each time a function is called, and is verified before returning from the function. They exit

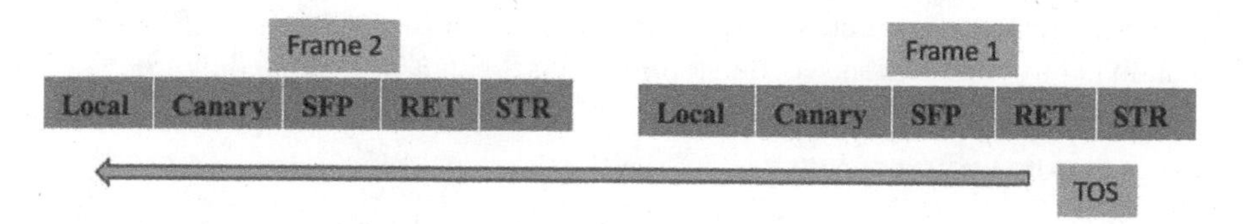

FIGURE 9.11 Stack containing a canary.

the program if a change in canary value is found. This randomness prevents most prediction attempts. It converts potential exploit into DoS.

- Terminator Canary – Most buffer-overflow attacks depend on string operations that end at string terminators. Four control characters included in this canary are able to terminate most string operations, leaving the overflow attempt harmless. This prevents the string function from copying beyond the terminator.
- Random XOR Canaries – Random canaries are XOR-scrambled using control data (frame pointer+return address+random number, etc.). In this way, if the canary or the control data is tampered with, the canary value changes and results in immediate program termination.

9.8.2 Pointguard

Pointguard is a mechanism for protecting programs against pointer corruption attacks. Some security attacks try to corrupt pointers stored in the address space of the program. To protect against such attacks, each time a pointer is initialized or modified, its value is, and decrypted before use. Encryption and decryption can be implemented by XOR'ing the pointer with a predetermined encryption key value, which could be pre-decided or randomly selected.

9.8.3 Libsafe

Libsafe is a mechanism that protects against the most commonly found buffer-overflow attacks. These overflow attacks enable the attackers to gain all types of privileges on the target system. Libsafe works as a shared library that captures and validates calls to vulnerable standard library functions. By inspecting the contents of the process stack and the function arguments, libsafe ensures that nothing can be overwritten on return addresses, thus providing defense against most of the buffer-overflow attacks commonly found.

The basic idea behind libsafe is to exchange vulnerable functions with safer alternatives. In the first step, libsafe first checks the safety of the called function. After it ensures that the function can be safely called. If yes, then it either calls the original function or executes functionally equivalent code. Otherwise, it displays warnings and terminates the process. As libsafe is incorporated as a shared library that is loaded into memory, it is able to replace its alternatives in place of the original functions. For Linux systems, the responsibility of loading the various program code and libraries into the memory is performed by the runtime loader. If libsafe is activated, the libsafe library is loaded into memory prior to the standard library. As the alternative libsafe functions have the same names as the original functions, the loader replaces the standard library functions with the libsafe functions. First, a safety check is done by most of the libsafe functions and then either the original function or a safer alternative is called. (e.g., snprintf()is called in place ofsprintf()) [24].

In [25], researchers proposed a novel technique to detect buffer-overflow attacks. The method intercepts all calls to vulnerable library functions, and an alternate version of the intercepted function implements the original functionality, ensuring that any buffer overflows will be limited within the boundaries of the current stack frame. The key idea is to automatically estimate a safe upper limit on the size of buffers. Compilers cannot provide this estimation as the size of the buffer may not be known at that time. The buffer size must be calculated after the start of the function that is using the buffer. It was realized that the local buffers could not extend beyond the end of the current stack frame. This realization permits the substitute function to restrict buffer writes within the scope of the estimated buffer size. Thus, the function's return address, which is placed on the stack, cannot be overwritten, and alteration of the control of the process cannot happen.

9.8.4 Control-Flow Integrity

CFI was originally proposed by Abadi et al. [26]. CFI defines a control-flow graph that specifies all possible paths a program may take during execution. It tries to stop control-flow hijacking by protecting against ROP. The strength of any CFI-based defense depends on how effectively the CFG is defined. If the CFG is perfectly constructed, then we can ensure only valid control flows are implemented by the program. However, perfectly constructing and maintaining CFG ahead of time requires a clear knowledge of the source code of the program, and incurs a very high performance cost. To ease the difficulty level and cost a weaker version of CFI can be used, but it guarantees weaker security.

CFI protects against control-hijacking attacks by ensuring that the control flow remains within the intended control-flow graph specified by the programmer. A unique ID is assigned to every legal instruction within the intended control flow, and checks are inserted before every control-flow instruction to ensure that only valid targets are allowed. Control-flow transfers can be direct or indirect. In the direct control flow, control is transferred to a fixed target, so checking can be done easily by verifying the control flow transfer. Whereas indirect transfers, such as function calls and returns, and indirect jumps, have a dynamic target address. As the attackers could control the dynamic target address due to some vulnerability, CFI prevents this by ensuring that its ID matches the list of known and allowable target IDs of the instruction. Every control flow that is defined in the CFG is permitted.

The CFI techniques are primarily classified as fine-grained and coarse-grained [27]. Both classifications comprise advantages and drawbacks in terms of control-flow security.

- Fine-Grained CFI uses a strict approach [28]. Labeling is one of the most common approaches to implementing fine-grained CFI. The edges of a control-flow transfer are labeled, and are checked for integrity on each control transfer. Control-flow transfer is preceded by a validation process to ensure that the transfer is passed to a valid destination with a valid label.
- Coarse-Grained CFI is referred to as loose CFI [28]. Coarse CFI differentiates between indirect branch instructions and imposes policies on each type. Loose CFI checks when a control-flow transfer initiates with a return

instruction so that the destination can be targeted right after the call instruction. This CFI approach is more relaxed and gives a desirable performance, and does not experience high-performance overhead; however.

9.8.5 Control Point Integrity

Code Pointer Integrity (CPI) banks on safeguarding the integrity of sensitive pointers to prevent control hijacking. As sensitive pointers belong to a superset of all pointers, CPI guarantees to provide robust security at a very fair and reasonable performance cost. CPI first performs a static analysis of the code to identify all sensitive pointers. It then saves metadata for inspecting the correctness of code pointers in a designated "safe region" of memory. The metadata contains the value of the pointer along with its lower and upper bounds. CPI heavily depends on the isolation of a restricted area that shares the space with the process being protected. CPI policies aim to prevent the control-hijacking attacks by guaranteeing the integrity of all sensitive pointers (e.g., function pointers) in a program. Levee [28] demonstrated the first practical implementation of CPI, introduced in 2014.

9.9 CONCLUSION

It is a never-ending affair: As new and effective mitigation techniques are offered, attackers are one step ahead and use existing vulnerabilities in the programming languages to perform a different type of control-hijacking attacks [29–35]. The year 2020 was affected by a cyber-attack pandemic along with COVID-19. The cyber-attacks that happened were ransomware, data breaches, healthcare attacks, and many more. Many cyber security professionals have predicted a heavily impacted 2021 as nearly all work has migrated to an online work-from-home culture. There will be more attacks on home computers and networks, if the system is unpatched and the architecture is weak. Moreover, since almost all the businesses are rushing to cloud-everything, this will also result in many challenges to the security, vulnerabilities, inappropriate configuration, and interruption of service.

REFERENCES

1. M. Payer. Control-Flow Hijacking: Are We Making Progress? ASIA CCS '17, Abu Dhabi, United Arab Emirates.
2. N. Burow, S. A. Carr, J. Nash, P. Larsen, M. Franz, S. Brunthaler, and M. Payer. Control-Flow Integrity: Precision, Security, and Performance. *ACM Computing Surveys*, 50(1), 2018.
3. L. Szekeres, M. Payer, L. Wei, D. Song, and R. Sekar. Eternal War in Memory. *IEEE Security and Privacy Magazine*, 2014.
4. C. Cowan, M. Barringer, S. Beattie, G. Kroah-Hartman, M. Frantzen, and J. Lokier. FormatGuard: Automatic Protection from Printf Format String Vulnerabilities. In *Proceedings of 10th USENIX Security Symposium*, August 2001.
5. Vendicator. StackShield GCC Compiler Patch. January 2001. http://www.angelfire.com /sk/stackshield.

6. D. Wheeler. Flawfinder. http://www.dwheeler.com/flawfinder.
7. J. Viega, J. T. Bloch, T. Kohno, and G. McGraw. ITS4: A Static Vulnerability Scanner for C and C++ code. In *Proceedings of the 16th Annual Computer Security Applications Conference*, December 2000.
8. C. Zhang, M. Niknami, K. Z. Chen, C. Song, Z. Chen, and D. Song. JITScope: Protecting Web Users from Control-Flow Hijacking Attacks. In *2015 IEEE Conference on Computer Communications (INFOCOM)*, 2015.
9. A. Smirnov, and T. Chiueh. DIRA: Automatic Detection, Identification, and Repair of Control-Hijacking Attacks. NSF awards ACI0234281, CCF-0342556, SCI-0401777, CNS-0410694 and CNS-0435373.
10. T. C. Chiueh, and F. H. Hsu. RAD: A Compile-Time Solution to Buffer Overflow Attacks. In *Proceedings of 21st International Conference on Distributed Computing Systems*, 2001.
11. J. Nazario. Project Pedantic – Source Code Analysis Tool(s). March 2002. http://pedantic.sourceforge.net.
12. Secure Software Solutions. Rough Auditing Tool for Security, RATS 2.1. http://www.securesw.com/rats.
13. H. Etoh. GCC Extensions for Protecting Applications from Stack-Smashing Attacks. June 2000. http://www.trl.ibm.com/projects/security/ssp.
14. P. Team. Non-Executable Pages Design and Implementation. http://pax.grsecurity.net/docs/noexec.txt.
15. Benjamin et al., Detection and Prevention of Stack Buffer Overflow Attacks. Communications of the ACM November 2005/vol. 48, no. 11.
16. Oenwall Project. http://www.openwall.com.
17. J. Ren, Z. Zheng, Q. Liu, Z. Wei, and H. Yan. A Buffer Overflow Prediction Approach Based on Software Metrics and Machine Learning. *Security and Communication Networks*, 2019.
18. Aleph One "Smashing the Stack for Fun and Profit" Edition 49 Phrack Magazine. November 1996.
19. Y. Younan, W. Joosen, and F. Piessens. Efficient Protection Against Heap-Based Buffer Overflows Without Resorting to Magic. In *8th International Conference on Information and Communications Security*, November 2006.
20. D. J. Day, Z. Zhao, and M. Ma. Detecting Return-to-libc Buffer Overflow Attacks Using Network Intrusion Detection Systems. ICDS 2010, St. Maarten, Netherlands Antilles.
21. N. Nagappan, and T. Ball. Static Analysis Tools as Early Indicators of Pre-Release Defect Density ICSE '05, May 15–21, 2005.
22. H. Shacham. The Geometry of Innocent Flesh on the Bone: Return into-libc without Function Calls (on the x86). In *Proceedings 14th ACM Conference Computer and Communications Security (CCS 07), ACM*, pp. 552–561, 2007.
23. A. Welekwe. Buffer Overflow Vulnerabilities and Attacks Explained. 2020.
24. D. Brumley, T. Chiueh, and R. Johnson. RICH: Automatically Protecting Against Integer-Based Vulnerabilities.
25. "tf8." Wu-Ftpd Remote Format String Stack Overwrite Vulnerability. http://www.Securityfocus.com/bid/1387, June 22, 2000.
26. M. Prasad, and T. Chiueh. A Binary Rewriting Defense against Stack based Buffer-Overflow Attacks.
27. H. Marco-Gisbert, and I. Ripoll. Address Space Layout Randomization Next Generation. *Applied Science Journal*, 2019, 9, 2928; doi:10.3390/app9142928.
28. H. Shacham et al., On the Effectiveness of Address-Space Randomization.
29. C. Cowan et al., StackGuard: Automatic Adaptive Detection and Prevention of Buffer-Overflow Attacks. In *7th USENIX Security Symposium San Antonio, Texas, January 26–29*, 1998.

30. T. Tsai, and N. Singh. Libsafe 2.0: Detection of Format String Vulnerability Exploits.
31. A. Baratloo, T. Tsai, and N. Sing. Libsafe: Protecting Critical Elements of Stacks.
32. M. Abadi et al., Control-Flow Integrity Principles, Implementations, and Applications CCS'05, November 7–11, 2005, Alexandria, Virginia, USA.
33. S. Sayeed, and H. Marco-Gisbert. Control-Flow Integrity: Attacks and Protections. *Applied Sciences* 9(20), 4229, 2019. https://doi.org/10.3390/app9204229.
34. I. Evans et al., Control jujutsu: On the Weakness of Control Flow Integrity In *Proceedings of ACM CCS*, 2015.
35. M. Muench et al., Taming Transactions: Towards Hardware-Assisted Control Flow Integrity using Transactional Memory. In *Proceedings of the 19th International Symposium on Research in Attacks, Intrusions and Defenses, France*, September 2016.

10 IoT Based Lightweight Cryptographic Schemes in Smart Healthcare

Dhirendra Siddharth, Priti Singh, and Dilip Kumar J Saini

CONTENTS

10.1 INTRODUCTION

Over the last few years, we've been working on cryptographic methods to increase the security of small devices such as RFID tags, embedded microcontrollers, and sensors. Digital assistants, smart city technologies, home automation, and healthcare applications all use these now-ubiquitous gadgets, which have limited computational power and storage space. Because these devices collect, store, and analyze so much sensitive data, users are concerned about their privacy and security. Furthermore, several of these products do not provide enough protection or employ proprietary, nonstandard security algorithms that can be reverse-engineered and cracked in

DOI: 10.1201/9781003229704-10

practice due to a lack of suitable cryptographic solutions that work successfully in these devices [1].

A lot of work has gone into creating new encryption algorithms that are optimized for limited devices in the last decade. These cryptographic algorithms are known as "lightweight" cryptographic algorithms. The term "lightweight" does not imply that the algorithms are inherently weak; rather, it relates to how simple they are to develop and how well they perform in limited environments [2].

The goal requirements and the optimum tradeoff between cost, performance, and security are defined by technology and applications in general. Anti-counterfeiting software frequently uses RFID tags with limited memory to identify and monitor retail items. Hardware-oriented algorithms that can be constructed fast are necessary in this instance [3]. Software-oriented solutions that require less memory are favored in smart home appliances with low-end computer processors.

As a result of the rapidly expanding need for Internet of Things (IoT) applications in medical, industrial, transportation, and other important applications, the security environment has altered radically. Unlike prior corporate apps that had free access to resources for processing security algorithms, business-level IoT applications are becoming increasingly vulnerable to assaults that target expanding networks of resource-constrained IoT devices. Organizations often install IoT devices that are operationally incapable of providing safety measures for preserving data stored and the flow of data and instructions over unsecured networks in their rush to respond to fast-developing IoT possibilities [4].

Of course, several factors impact how far designers must go to ensure a system's security. Each application has its unique set of dangers, demanding a comprehensive assessment of the threats' potential consequences. Because linked devices face a unique set of risks, each IoT device will need to integrate at least some basic security features.

To some, putting sufficient security safeguards in place for a basic IoT device may appear over-engineered, yet even unsecured temperature sensor devices can provide valuable network access. IoT security is a recurring worry due to the combination of extensive connectivity provided by IoT applications and resource-constrained hardware that enables these apps. Even with IoT device designs that have the processing capability to run cryptographic algorithms in software, apps can be vulnerable owing to minor implementation issues [5].

10.2 WHAT IS IoT?

The Internet of Things is a term that refers to a group of internet-connected devices. When you hear "Internet of Things," you might think of laptops or smart TVs, but the term actually refers to a lot more. Consider the copy machines, refrigerators, and coffee makers that have never been connected to the internet in the breakroom. Any device that can connect to the internet, including odd devices, is referred to as the Internet of Things. Almost everything with an on/off switch now has the ability to connect to the internet, making it a part of the Internet of Things [6].

10.2.1 Some of the Most Common IoT Workplace Applications

According to studies, IoT devices can help businesses operate more efficiently. According to Gartner, the top IoT benefits for businesses include increased employee productivity, remote monitoring, and simplified operations.

But how does the Internet of Things play out in the workplace? Here are a few instances of IoT connectivity in the workplace; however, each organization is different:

- Siri and Alexa can open applications that let you take notes, set reminders, check your calendar, and send emails.
- Smart thermostats and lights turn on and off as required to conserve energy.
- A company leader may unlock a smart lock door with his or her smartphone.
- CCTV cameras have the ability to transmit footage over the internet.
- Printers with connected sensors detect low ink levels and immediately place an order for additional ink.

10.2.2 What is IoT Security

In terms of increasing business productivity and connectivity, IoT devices have a lot of promise. They do, however, introduce additional issues, such as safeguarding sensitive data sent over the internet, avoiding device sabotage, and stopping IoT devices from becoming part of criminal botnets. Security must be a primary consideration when deploying or allowing Internet of Things devices on a company's network. Many businesses are unaware of the number of devices connected to their networks. Unmanaged network approaches, such as smartwatches and cellphones, as well as testing and other equipment, provide a significant security risk. This paper explains Internet of Things security, looks at the most pressing IoT security concerns now facing businesses, and discusses how to keep IoT devices up to date [7].

The term "Internet of Things" describes the millions of non-traditional computer devices that can transmit and receive data over the internet. From soil monitoring to internet-connected refrigerators to intelligent traffic lights, these technologies are becoming more prevalent. They all have access to sensitive data and have the ability to monitor and/or change essential systems.

In some respects, IoT security is similar to network security, but the vulnerability of data gathered by IoT devices and the apps they control changes the dynamic. An IoT device capable of shutting off a power supply or recording footage of a home's interior requires a higher level of protection than standard monitoring [8].

10.2.3 Managed Internet of Things Devices

Because IoT networks are comprised of numerous levels, each of which must be safeguarded, IoT security is extremely challenging to implement. Software operating on IoT devices, according to IT security experts, must be secure and regularly updated. To combat faults in the apps they use to engage with one another, IoT devices must overcome difficulties in the applications they use to communicate with one another. Intrusions into IoT networks must be identified as well. Finally, whether saved on IoT devices or sent to a cloud server, IoT data must be kept safe.

Some of these security procedures can be managed from a single point of management. For example, a security operations center might handle IoT equipment identification and software upgrades. Other areas of IoT security that require a variety of solutions include bug testing apps and assuring information security in transit and at rest [9].

Different groups are in charge of various elements of IoT device management.

- IT is in charge of equipment deployment and administration.
- The security team is in charge of keeping an eye on vulnerabilities and developing secure IoT solutions.
- Updates, changing default passwords, and other activities are also the responsibility of end-users.

While, in most cases, this separation of duties for IoT equipment administration is inescapable, it complicates IoT security by necessitating coordination among various stakeholders to make sure that the most excellent practices are followed and maintained.

10.2.4 Security Issues With IoT

As previously stated, IoT devices include security weaknesses and vulnerabilities that are not necessarily present in traditional hardware.

Updates are not trustworthy

IoT devices are frequently left unpatched, leaving them vulnerable to previously unknown security issues. For starters, IoT devices are tiny and inconspicuous. It's all too simple for companies to deploy IoT devices and then forget about them once a large number have been deployed. Furthermore, many IoT systems rely on users to maintain their systems current, which many users overlook or ignore.

"Shadow" IoT Devices

A key IoT security issue is the possibility of "shadow" devices, or equipment linked to an IoT system but not permitted or understood by the network owner. Shadow devices may be introduced into the network by unaware users, such as an employee who brings an IoT temperature monitor to work. Malicious actors may utilize them to industrialize intelligence through the use of smart televisions or unprotected forum phones [10].

A substantial proportion of equipment in installations is unlabeled or unauthorized, according to a new whitepaper titled Rise of the Machinery: 2020 Enterprise of Things Adoption and Risks Report (12–15%). The most noteworthy incident included a Tesla connecting to a medical system; after a lengthy investigation, security officials discovered that the Tesla belonged to a physician who connected to the internet from his parking garage car.

Implementation of standards

There are no standardized design principles for IoT devices, software, or data communication, and there is no standardized IoT API. Instead, a slew of competing IoT hardware and software solutions is continuously emerging.

Protecting IoT devices from a security standpoint is more difficult since there are several factors at play. There is no one protection solution that can save from harm all Internet of Things devices and networks from all threats.

API Vulnerabilities

API vulnerabilities are a big security concern since sending data over the network via API is such an important part of what IoT devices perform. Attackers can employ man-in-the-middle attacks to steal data or take control of a device to launch DDoS attacks if an API vulnerability is identified.

Because there is no one set of IoT Application Programming Interfaces to monitor, there is no single collection of IoT Application Programming Interface issues to be aware of. Instead, you could use one of the hundreds of IoT APIs offered by various manufacturers, or you could develop your own. Security teams should be aware of any possible dangers while utilizing APIs [11].

10.2.5 Threats to IoT Security Are Present in Businesses.

Any company that uses an IoT device is subject to the above-mentioned IoT security issues. The risks are particularly serious in some organizations because of the potential implications of the sensitivity of data collected by IoT devices:

Hospitality: Despite the fact that 70% of hospitality companies have adopted Internet of Things projects, the security risks offered by IoT are a big worry for them, owing to the risk of reputational damage if an attack happens.

Healthcare: IoT devices gather medical information and, in certain circumstances, may be placed into human bodies. In this circumstance, a security breach may have disastrous consequences.

Manufacturing: If a manufacturer's IoT network is compromised, it might result in substantial downtime and financial loss.

Government: Attackers may gain access to government IoT infrastructure in order to get privileged information or disrupt vital services when authorities rely on smart objects to collect data or operate physical infrastructures such as dams or motorways.

Transportation: A cyber-attack on Internet of Things devices may quickly destabilize smart cities that depend on them. If IoT devices are utilized to drive buses or pilot planes, their security flaws might have serious consequences.

Retail: The use of IoT devices in retail might aid in theft prevention, inventory management, and other tasks. Hackers might use unprotected IoT devices to steal personal information or disrupt essential corporate processes, on the other hand.

10.3 LIGHTWEIGHT CRYPTOGRAPHY

Lightweight cryptography is a term used to describe encryption for devices with limited resources, such as RFID tags and wireless sensor networks. Lightweight cryptography's objective is to enable a wide range of contemporary applications, including smart meters, car security systems, wireless patient monitoring systems, Intelligent Transportation Systems (ITS), and the IoT.

As a result, lightweight cryptography may be utilized on a wide range of devices. They may communicate via wired or wireless networks and use a wide range of hardware and software. Electromagnetic induction or a battery, which might be disposable or rechargeable, can power wireless gadgets [12]. As a result, although low latency is important for some applications, energy consumption is important for others. It's likely that a limiting issue is the size of the hardware area or software code, or that there's only so much RAM accessible. The majority of the time, a combination of the aforementioned requirements must be met.

It's worth noting that "lightweight cryptography" doesn't always mean "poor encryption." Short key or block sizes, for example, have been marketed as "lightweight" algorithms, indicating that they are significantly less powerful than traditional algorithms. We are adamantly opposed to the employment of such algorithms, and we shall refute common arguments [13].

10.4 SYMMETRIC KEY CRYPTOGRAPHY

Starting small is the key to achieving symmetry. In contrast to asymmetric key cryptography, which uses distinct encryption and decryption keys, cryptography is any crypto technique that encrypts or decrypts text using a single key. Symmetric encryption is favored over asymmetric encryption when transferring large volumes of data since it is more efficient. Asymmetric encryption is utilized in a number of circumstances since it is difficult to produce public keys between two parties using just symmetrical encryption methods. DES, AES, and 3DES are examples of symmetric key cryptography. A shared encryption key is created using the key exchange techniques elliptic curve (EC), Diffie-Hellman (DH), and RSA [14].

10.5 PUBLIC-KEY CRYPTOGRAPHY

The process of encrypting data using two distinct keys is known as public-key cryptography (also known as public-key encryption or asymmetric cryptography). The public key is one of the keys that anybody may use. The other key is the private key, which is unique to each individual. Anyone with the recipient's public key can encrypt a message, but only the recipient's private key can decode it. This ensures that only the intended recipient sees the communication [15].

10.6 LIGHTWEIGHT CRYPTOGRAPHY WITH HIGH SECURITY

The National Institute of Standards and Technology (NIST) has started working on a cryptography concept that has gotten little notice so far. According to NIST, the technique is referred to as "lightweight cryptography," and it seeks to "develop crypto algorithm principles that can work within the limits of a simple electrical device." The statement was issued by the NIST in reaction to the rapidly expanding Internet of Things, which is a network of monitors, cameras, sensors, and other devices that collaborate to build smart services. The Internet of Things includes self-driving cars, smart energy systems, and other technical advancements aimed at increasing efficiency. Without the Internet of Things, none of these systems could function.

According to the NIST, the millions of electrical gadgets that make up the Internet of Things are unable to handle existing encryption techniques due to their tiny size and simplicity. Lightweight cryptography will need far fewer resources from systems, allowing them to perform fundamental tasks in a fraction of the time [16]. The cost of installation will become prohibitive if firms choose expensive, heavy-weight solutions for every single IoT device. As a result, lightweight cryptography would be more suited to protecting the sensitive data flows that occur on the Internet of Things every second.

Lightweight cryptography is one of the cryptographic solutions on the horizon. As the Internet of Things expands and projects like self-driving vehicles and smart cities arise, lightweight encryption will undoubtedly become a necessary part of daily urban life. Return to the future for the most straightforward information on this critical IoT network security initiative [17].

10.7 LIGHTWEIGHT CRYPTOGRAPHY'S POWER AND ENERGY

The term "lightweight cryptography" refers to algorithms that provide security while consuming fewer resources, such as electricity or energy. The amount of power consumed by an algorithm is determined by its design architecture and implementation. As a result, different algorithm implementation designs use different amounts of energy and power. Finally, power-saving measures might be included in a hardware implementation for certain architecture. In this part, we'll discuss power use at these three levels, taking into consideration both stream and block ciphers.

SKC ciphers that create a keystream that is XORed with the plaintext to form the ciphertext are known as stream ciphers. Serial shift registers and a pseudorandom sequence generator with a random seed value are used to produce the secret key and an initialization vector. Block ciphers, on the other hand, work with huge blocks of digits and a predetermined transformation, mixing plain text and the key with basic operations like substitutions and permutations in several rounds. Because the algorithmic, architectural, and implementation components of stream and block ciphers are so distinct, both forms of ciphers require independent investigation and attention [18].

10.7.1 Power Consumption Issues at an Algorithmic Level

Stream Cipher Algorithms

Shift registers with linear and non-linear feedback are commonly used to generate pseudorandom bit sequences. As a result, the number of shift registers and the complexity of the feedback function have a significant impact on the amount of power consumed by a stream cipher.

Block Cipher Algorithms

The algorithms for block ciphers are more complicated than those for stream ciphers. Many block ciphers have been proposed, and they may be divided into two categories: substitution-permutation networks and Feistel networks. The cipher serves two functions in an SPN: confusion and dispersion. A layer of Substitution boxes, or S-boxes, is used to create confusion, which is essentially a permutation of a tiny subset of data. Diffusion occurs when the entire space, which is usually linear, gets permuted. The data block is split into two equal halves and encrypted in multiple rounds with FN ciphers, allowing permutation and combinations based on the primary function or key [19].

10.7.2 Factors Related to Power Consumption at the Architecture Level

Stream cipher architectures

As previously stated, stream ciphers have an internal structure that is defined by the algorithm. Multi-bit stream ciphers, or stream ciphers that output several bits in a single clock cycle, are the most efficient way to cut down on power consumption and, more precisely, energy per bit. These techniques maintain the internal state size while increasing the number of bits required for feedback. As a consequence, during each clock cycle, n bits are generated in parallel [20].

Block ciphers architectures

Block ciphers iterate through a sequence of operations, completing a number of rounds in a parallel or serial manner. Each round of a clock cycle can be performed using block cipher implementations, also known as rolling implementations. In contrast, unrolled versions can complete many rounds in a single clock cycle. Rolling implementations utilize fewer resources in general, but it takes longer to perform an encryption operation due to the extra clock cycles. The findings may differ in terms of power usage or energy per bit [20].

10.8 IDENTIFYING THE BENEFITS AND DEMERITS OF LIGHTWEIGHT CIPHERS

Since lightweight algorithms were created to achieve certain goals, their advantages and disadvantages are obvious. Lightweight algorithms are extremely fast to operate and "unpretentious" to the environment in which they will be utilized for work because of their extremely low resource and power consumption. It also saves money on the cost of developing future lightweight algorithms.

Lightweight algorithms, on the other hand, are intended to handle small amounts of data and hence have limited bandwidth. Light ciphers were designed primarily for hardware implementation rather than soft implementation, as seen by the inclusion of limitations in 1000 GE. When more GE is required, such as to implement additional S-boxes in the block cipher algorithm or to use a long key, lightweight algorithm writers face a significant difficulty [21].

Additionally, current algorithms are difficult to adapt to the demands of lightweight cryptography because many of them have lost significant resistance owing to their restricted computational resources. The development of lightweight algorithms is sluggish, despite the existence of ISO/IEC 29192, an encryption lite series standard. Furthermore, the number of successful attacks on existing lightweight cryptography systems is steadily increasing [22].

10.9 CONCLUSION

Depending on your IoT application and where you are in the IoT ecosystem, you may use a variety of IoT security approaches. For example, IoT device manufacturers should prioritize security by developing secure hardware, making it tamper-proof, providing firmware updates/patches, guaranteeing safe upgrades, and doing dynamic testing. We talked about IoT scenarios with limited resources and possibly lightweight cryptography solutions. Lightweight cryptography considers key management functions and operations in real-world applications in addition to key management functions and operations. In key exchanges that employ public-key encryption, we're considering utilizing lightweight cryptography. In the future, we intend to continue researching cryptographic technologies in order to contribute to more secure IoT networks.

REFERENCES

1. S Al Salami, J Baek, K Salah, and E Damiani, "Lightweight Encryption for Smart Home," In: *Proceeding of 2016 11th International Conference on Availability, Reliability and Security (ARES)*, IEEE, pp. 382–388, 2016.
2. S Babbage, and M Dodd, "The MICKEY Stream Ciphers," In: *Proceeding of New Stream Cipher Designs*, Springer, Berlin, pp. 191–209, 2008.
3. G Bansod, N Raval, and N Pisharoty, "Implementation of a New Lightweight Encryption Design for Embedded Security," *IEEE Trans. Inf. Forens. Sec.*, vol. 10, no. 1, pp. 142–151, 2015.
4. R. Want, B. N. Schilit, and S. Jenson, "Enabling the Internet of Things," *IEEE Comput.*, vol. 48, no. 1, pp. 28–35, Jan. 2015.
5. L. Baresi, L. Mottola, and S. Dustdar, "Building Software for the Internet of Things," *IEEE Internet Comput.*, vol. 19, no. 2, pp. 6–8, March/April 2015.
6. Y. Qin et al., "When Things Matter: A Survey on Data-Centric Internet of Things," *J. Netw. Comput. Appl.*, vol. 64, pp. 137–153, February 2016.
7. Z. Zhang, M. C. Y. Cho, C. Wang, C. Hsu, C. Chen, and S. Shieh, "IoT Security: Ongoing Challenges and Research Opportunities," In: *2014 IEEE 7th International Conference on Service-Oriented Computing and Applications*, pp. 230–234, 2014. https://doi.org/10.1109/SOCA.2014.58.

8. W. Shang, Q. Ding, A. Marianantoni, J. Burke, and L. Zhang, "Securing Building Management Systems Using Named Data Networking," In: *IEEE Network Special Issue on Information-Centric Networking*, April 2014.
9. S. Raza, H. Shafagh, K. Hewage, R. Hummen, and T. Voigt, "Lithe: Lightweight Secure CoAP for the Internet of Things," *IEEE Sens. J.*, vol. 13, no. 10, 3711–3720. 2013.
10. P. H. Cole, and D. C. Ranasinghe, *Networked RFID systems and lightweight cryptography*. Springer, London, 2008.
11. A. Costin, J. Zaddach, A. Francillon, and D. Balzarotti. "A Large Scale Analysis of the Security of Embedded Firmwares," In: *USENIX Security Symposium*, August 2014.
12. K. McKay, L. Bassham, M. Sonmez Turan, and N. Mouha, "Report on Lightweight Cryptography," Jul. 2017 [online]. Available: https://csrc.nist.gov/publications/detail/nistir/8114/final.
13. S. Singh, P. K. Sharma, S. Y. Moon, and J. H. Park, "Advanced Lightweight Encryption Algorithms for IoT Devices: Survey Challenges and Solutions," *J. Ambient Intell. Human Comput.*, pp. 1–18, 2017 [online]. https://doi.org/10.1007/s12652-017-0494-4.
14. H. Delfs, and H. Knebl, "Symmetric-Key Cryptography," In: *Introduction to Cryptography. Information Security and Cryptography*. Springer, Berlin, Heidelberg. https://doi.org/10.1007/978-3-662-47974-2_2.
15. W. Diffie, "The First Ten Years of Public-Key Cryptography," *Proceedings of the IEEE*, vol. 76, no. 5, pp. 560–577, May 1988. https://doi.org/10.1109/5.4442.
16. X. Yao, Z. Chen, and Y. Tian, "A Lightweight Attribute-Based Encryption Scheme for the Internet of Things," *Future Gener. Comput. Syst.*, vol. 49, pp. 104–112, 2015. https://doi.org/10.1016/j.future.2014.10.010.
17. I. Bhardwaj, A. Kumar, and M. Bansal, "A Review on Lightweight Cryptography Algorithms for Data Security and Authentication in IoTs," In: *2017 4th International Conference on Signal Processing, Computing and Control (ISPCC)*, pp. 504–509, 2017. https://doi.org/10.1109/ISPCC.2017.8269731.
18. K. Zhao, and L. Ge, "A Survey on the Internet of Things Security," In: *Ninth International Conference on Computational Intelligence and Security*, pp. 663–667, December 2013.
19. N. S. Bhagya, K. Murad, and H. Kijun, "Towards Sustainable Smart Cities: A Review of Trends, Architectures, Components, and Open Challenges in Smart Cities," *Sustain. Cities Soc.*, vol. 38, pp. 697–713, 2018.
20. B. J. Mohd, and T. Hayajneh, "Lightweight Block Ciphers for IoT: Energy Optimization and Survivability Techniques", *IEEE Acc.*, vol. 6, pp. 35966–35978, 2018. https://doi.org/10.1109/ACCESS.2018.2848586.
21. G. Leander, C. Paar, A. Poschmann, and K. Schramm, "New Lightweight Des Variants," *Proc. Int. Workshop Fast Softw. Encryption*, 4593, pp. 196–210, 2007.
22. D. J. Wheeler, and R. M. Needham, "TEA a Tiny Encryption Algorithm," *Proc. Int. Workshop Fast Softw. Encryption*, 1008, pp. 363–366, 1994.

Index